T0252922

ROCK MAGNETIC CYCLOSTRATIGRAPHY

New Analytical Methods in Earth and Environmental Science

A new e-book series from Wiley-Blackwell

Because of the plethora of analytical techniques now available, and the acceleration of technological advance, many earth scientists find it difficult to know where to turn for reliable information on the latest tools at their disposal, and may lack the expertise to assess the relative strengths or potential limitations of a particular technique. This new series addresses these difficulties, and by virtue of its comprehensive and up-to-date coverage, provides a trusted resource for researchers, advanced students and applied earth scientists wishing to familiarise themselves with emerging techniques in their field.

Authors will be encouraged to reach out beyond their immediate speciality to the wider earth science community, and to regularly update their contributions in the light of new developments.

Written by leading international figures, the volumes in the series will typically be 75–200 pages (30,000 to 60,000 words) in length – longer than a typical review article, but shorter than a normal book. Volumes in the series will deal with:

- the elucidation and evaluation of new analytical, numerical modelling, imaging or measurement tools/techniques that are expected to have, or
- are already having, a major impact on the subject; new applications of established techniques;
- interdisciplinary applications using novel combinations of techniques.

All titles in this series are available in a variety of full-colour, searchable e-book formats and may include additional features such as DOI linking, high resolution graphics and video. See individual books for details.

Series Editors

Kurt Konhauser, University of Alberta (biogeosciences)
Simon Turner, Macquarie University (magmatic geochemistry)
Arjun Heimsath, Arizona State University (earth-surface processes)
Peter Ryan, Middlebury College (environmental/low T geochemistry)
Mark Everett, Texas A&M (applied geophysics)

ROCK MAGNETIC CYCLOSTRATIGRAPHY

KENNETH P. KODAMA
Department of Earth and Environmental Sciences, Lehigh University, Bethlehem, PA, USA

LINDA A. HINNOV
Department of Earth and Planetary Sciences, Johns Hopkins University, Baltimore, MD, USA

WILEY Blackwell

This edition first published 2015 © 2015 by John Wiley & Sons, Ltd.

Registered Office
John Wiley & Sons, Ltd, The Atrium, Southern Gate, Chichester, West Sussex, PO19 8SQ, UK

Editorial Offices
9600 Garsington Road, Oxford, OX4 2DQ, UK
The Atrium, Southern Gate, Chichester, West Sussex, PO19 8SQ, UK
111 River Street, Hoboken, NJ 07030-5774, USA

For details of our global editorial offices, for customer services and for information about how to apply for permission to reuse the copyright material in this book please see our website at www.wiley.com/wiley-blackwell.

The right of the author to be identified as the author of this work has been asserted in accordance with the UK Copyright, Designs and Patents Act 1988.

All rights reserved. No part of this publication may be reproduced, stored in a retrieval system, or transmitted, in any form or by any means, electronic, mechanical, photocopying, recording or otherwise, except as permitted by the UK Copyright, Designs and Patents Act 1988, without the prior permission of the publisher.

Designations used by companies to distinguish their products are often claimed as trademarks. All brand names and product names used in this book are trade names, service marks, trademarks or registered trademarks of their respective owners. The publisher is not associated with any product or vendor mentioned in this book.

Limit of Liability/Disclaimer of Warranty: While the publisher and author(s) have used their best efforts in preparing this book, they make no representations or warranties with respect to the accuracy or completeness of the contents of this book and specifically disclaim any implied warranties of merchantability or fitness for a particular purpose. It is sold on the understanding that the publisher is not engaged in rendering professional services and neither the publisher nor the author shall be liable for damages arising herefrom. If professional advice or other expert assistance is required, the services of a competent professional should be sought.

Library of Congress Cataloging-in-Publication Data

Kodama, Kenneth P.
Rock magnetic cyclostratigraphy / Kenneth P. Kodama, Linda A. Hinnov.
    pages cm
  Includes index.
  ISBN 978-1-118-56128-7 (cloth)
1. Cyclostratigraphy. 2. Paleomagnetism. 3. Geochronometry. I. Hinnov, L. A. (Linda Alide) II. Title.
  QE651.5.K63 2015
  551.7'01–dc23

                                                  2014025855

A catalogue record for this book is available from the British Library.

Wiley also publishes its books in a variety of electronic formats. Some content that appears in print may not be available in electronic books.

Cover image: *Moonscape*. Original painting by Anna Kodama

Set in 10/12.5pt Minion by SPi Publisher Services, Pondicherry, India

1   2015

Contents

1

Introduction

Abstract: Rock magnetic cyclostratigraphy is a new technique that allows a high-resolution chronostratigraphy to be assigned to a sequence of sedimentary rocks. Concentration variations of magnetic minerals in a sedimentary rock can be tied to astronomically forced global climate cycles with little or no facies interpretation needed. The rock magnetic measurements are nondestructive, relatively quick, and inexpensive. This chapter outlines the basic steps of a rock magnetic cyclostratigraphy study and serves as an introduction to the monograph.

1.1 Rock Magnetic Cyclostratigraphy

The purpose of this monograph is to provide an overview and the practical "how to" for a relatively new technique that can yield high-resolution chronostratigraphy for sequences of sedimentary rocks. Rock magnetic cyclostratigraphy is the result of the merging of environmental magnetism, in which rock magnetic measurements can detect past environmental conditions, and cyclostratigraphy, in which cyclic variations of lithology or a rock's physical properties are tied to orbitally forced changes in global climate. Orbitally forced cyclic variations in the lithology of sedimentary sequences has been an important research focus for stratigraphers since Hays et al.'s (1976) pioneering study of Late Pleistocene marine sediments. The main reason for this intense interest is that if lithologic variations can be tied to the well-known cyclic variations of solar insolation at periods of ~20 kyr, ~40 kyr, ~100 kyr, and 405 kyr, a detailed and high-resolution chronostratigraphy can be established for the rocks, even at distant times in Earth's history.

Lithologic cyclostratigraphy relies on identifying facies changes in a rock sequence and interpreting them as indicators of cyclic variations in the rock's depositional environment. These cyclic variations are then tied to astronomically forced climate change. Deep-sea cyclostratigraphy has been

Rock Magnetic Cyclostratigraphy, First Edition. Kenneth P. Kodama and Linda A. Hinnov.
© 2015 John Wiley & Sons, Ltd. Published 2015 by John Wiley & Sons, Ltd.

vital in supplying pristine records of astronomically forced signals from the Cenozoic and Late Mesozoic eras. For earlier times, however, pelagic marine organisms had not yet evolved in sufficient "rock-forming" numbers. For the early Mesozoic and earlier times, researchers must rely on shallow marine, hemipelagic, and continental cyclostratigraphy for astronomically forced paleoclimate data. While continental facies preserve high fidelity records of astronomical forcing, e.g., the Newark Basin lacustrine rocks (Olsen & Kent 1996), such facies are in short supply compared with the marine record. Shallow-marine cyclostratigraphy, principally from carbonate-rich peritidal facies, is the main source of astronomical forcing and global climate change data prior to the Jurassic Period (>200 million years ago). However, in any lithologic, facies-based cyclostratigraphic study, the work always involves interpretation, both in the identification of a given facies and in the interpretation of what that facies indicates about the depositional environment.

To advance the study of cyclostratigraphy, stratigraphers have searched for techniques that could provide stable and well-behaved paleoclimatic or paleoenvironmental proxies at high resolution and could be collected over reasonably thick sedimentary sequences. The Holy Grail would be a simple, low cost and fairly quick measurement that would be amenable to time series analysis and require minimal interpretation. For instance, in the recognition of astronomically forced cycles in the Late Triassic lake sediments of the Newark Basin, Olsen & Kent (1996) assigned depth ranks to quantify the depositional environment interpreted from the facies changes in the rocks. With the construction of a rock magnetic time series, the facies/depositional environment interpretation could be short-circuited, and the rock magnetics would directly quantify the paleoenvironmental/paleoclimate change.

Rock magnetics and rock magnetic cyclostratigraphy can fulfill many of these needs. Rock magnetic parameters are used in the subdiscipline of environmental magnetism to detect the ancient depositional environment. Rock magnetic parameters can measure variations in the concentration, particle size, and mineralogy of magnetic minerals in a sedimentary rock. These measurements are relatively quick and, therefore, inexpensive, so 1000s of samples can be collected and measured for a rock magnetic cyclostratigraphic study to document magnetic variations at high resolution. The measurements are also nondestructive, so the samples can be retained for other nonmagnetic measurements and examination. The variations in magnetic mineral concentration and particle size can be tied to changes in the depositional environment and hence to changes in paleoclimate or paleoenvironment. Since magnetic minerals in Earth's crust all contain iron, either as oxides, oxyhydroxides, or sulfides, and iron is the fourth most common element in the crust, magnetic minerals can sensitively delineate the cycling of this ubiquitous element through Earth's atmosphere, biosphere, lithosphere, and hydrosphere. Furthermore, very small concentrations of magnetic minerals (<0.01%) are easily and accurately measured with modern superconducting rock magnetometers, making rock magnetic measurements very sensitive measures of paleoenvironmental conditions. In one of the case studies presented in Chapter 6, the sensitivity of rock magnetics to paleoclimatic variations will be demonstrated by a study of the Cretaceous Cupido Formation from Mexico in which rock magnetics can detect astronomically forced cycles, even though the repeating, shallowing-upward facies cannot.

Rock magnetic parameters have been successful measures of glacial–interglacial cycles in loess, mainly in the Chinese Loess Plateau, but also in Eastern Europe and Alaska (summarized in Evans & Heller 2003). Rock magnetic measurements of European maar lake sediments have also detected glacial–interglacial climate cycles. Susceptibility variations from Lac Du Bouchet in France have been directly correlated to $\delta^{18}O$ records of glacial–interglacial cycles from the Pacific and Indian Oceans and Greenland ice cores (Heller et al. 1998). Terrigenous input into the northwestern Indian Ocean can be tracked by magnetic susceptibility, and the cyclic variations in susceptibility can be directly correlated to astronomical calculations for northern hemisphere insolation (deMenocal & Bloemendal 1995). Susceptibility variations have also detected changes in paleoclimate in Eocene marine sediments off Antarctica (Sagnotti et al. 1998). Various studies of North Atlantic marine sediments have used rock magnetics to study deglaciation (Stoner et al. 1995) and North Atlantic Deep Water circulation (Kissel et al. 1999). These examples show that rock magnetic parameters that measure a quantity as simple as the concentration of magnetic minerals in sediment can easily detect changes as profound as global paleoclimate.

Conducting a rock magnetic cyclostratigraphic study of a sedimentary sequence is fairly straightforward. Most rock magnetic cyclostratigraphic studies measure variations in the concentration of a depositional magnetic mineral in a sequence of rocks. Magnetite (Fe_3O_4) is, in most cases, a primary, depositional magnetic mineral. Therefore, erosional, transport, and depositional processes as well as the depositional environment affect its concentration, making magnetite the preferred target of cyclostratigraphic studies. Furthermore, magnetite has a relatively low magnetic coercivity (for magnetic hardness, see Chapter 2) and its concentration is easily measured by applying an anhysteretic remanent magnetization (ARM) (for more details, see Chapter 2) to the cyclostratigraphy samples. ARM, as will be shown in Chapter 2, also allows the researcher to target the concentration variations of only one magnetic mineral (magnetite) in the rock compared to the multiple mineral sources for magnetic susceptibility, so the interpretation of any rock magnetic cycles recorded by an ARM will be straightforward. However, as shown in the case studies presented in Chapter 6, other rock parameters can be used with equal success for identifying astronomically forced cycles in a sedimentary sequence. The rock magnetic parameters used must be chosen on a case-by-case basis.

1.2 Basic Steps of a Rock Magnetic Cyclostratigraphy Study

The steps to a rock magnetic cyclostratigraphy are summarized in Figure 1.1. The first step in conducting a rock magnetic cyclostratigraphy is to select the stratigraphic section for study and estimate its sediment accumulation rate, so that the correct sampling interval can be chosen. The frequencies that can be detected by the time series analysis are limited by the Nyquist frequency. The shortest cycle that can be observed by time series analysis must be sampled at least twice per cycle; therefore, if precession (nominally a 20 kyr period) is to be captured, the rocks should be sampled at least once every 10 kyr of stratigraphic thickness. In some cases, previous work, either biostratigraphy, magnetostratigraphy, or geochronology of ash layers, can be used to calculate the sediment accumulation rate. In most cases, though, only an estimate of the sediment accumulation rate, based on the rock's lithology and depositional environment, is available. Sadler's (1981) comprehensive study of sedimentary record completeness can be an important source of these estimates. However, sampling precession twice per cycle is only a bare minimum. Aliasing could occur if shorter cycles present in the data are undersampled, producing apparent longer period cycles which are only an artifact of the sampling interval. It may be better to target possible precession cycles with three or four samples per cycle.

Once the sampling interval is selected, unoriented rock samples, ~4–5 cm in size, are collected throughout the section. If the modulation of precession (~20 kyr) by short (~100 kyr) and long (405 kyr) eccentricity is the desired target for detection, then the sampling interval is at least a 10 kyr stratigraphic thickness of sediment (ideally 5–7 kyr), and the stratigraphic thickness of the sampled interval should be long enough to record about six repetitions of the longest period cycle (Weedon 2003). Furthermore, the longer the series, the better the bandwidth resolution, i.e., the narrower the spectral peaks and hence the better one can resolve the frequency of individual cycles. If long eccentricity is the longest period targeted for detection, then the section should be at least ~2–2.5 million years long. For typical hemipelagic marine sediments with sediment accumulation rates of about 10 cm/kyr (Sadler 1981), these requirements would mean a sampling interval of about 0.5–1 m and a section thickness of at least 200–250 m, generating a minimum of 200–250 samples. For best results, at least three or four samples should be collected for every precessional cycle, increasing the number of samples to 500 samples for the 2- to 2.5-million-years long section. The work needed to collect this large number of samples is offset somewhat because the samples do not need to be oriented, as would standard paleomagnetic samples, because only the intensity of the sample is measured, not the direction of its magnetization. Sampling a 250 m section at 75 cm intervals would take about 3–4 days in the field.

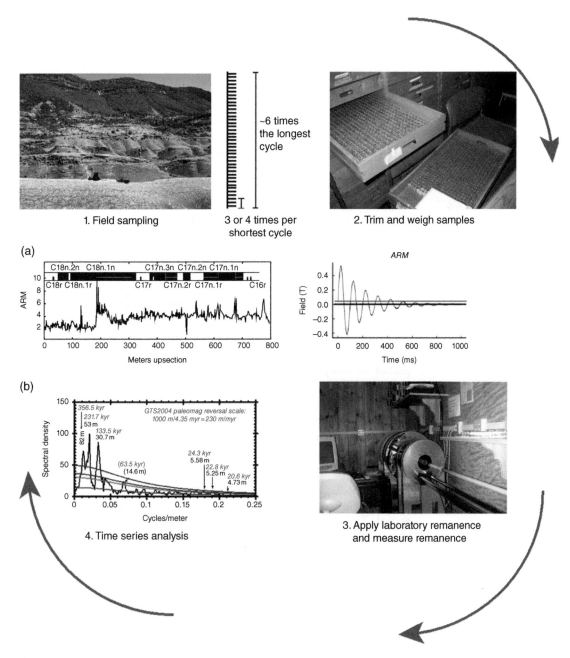

Figure 1.1 Steps to a rock magnetic cyclostratigraphy. Step 1: The field sampling picture is the Eocene Arguis Formation in the Spanish Pyrenees. Step 2: Cretaceous Cupido Formation limestone samples in 8 cm³ plastic sampling boxes. Step 3: Diagram of alternating and direct fields during application of an ARM. Lehigh University 2G Enterprises superconducting rock magnetometer. Step 4: (a) Arguis Formation ARM series and magnetostratigraphy developed for the Arguis Formation. (b) **Multitaper method (MTM)** power spectrum of the Arguis Formation ARM time series. Source: Kodama, Anastasio, Newton, Pares & L. A. Hinnov 2010. Reproduced with permission of John Wiley & Sons, Inc.

After the samples are brought to the laboratory, they must be trimmed to fit into standard paleomagnetic/rock magnetic plastic sample boxes (nominally 2 cm × 2 cm × 2 cm in size). The samples should be weighed because the magnetic measurements of intensity must be normalized by sample mass for an accurate cyclostratigraphy. Alternatively, small diameter (~11 mm) 15 mm long cores can be drilled from the samples. After the samples are trimmed or cored, placed in sample boxes and weighed, they are ready for rock magnetic analysis.

If ARM is the rock magnetic parameter selected for the cyclostratigraphy, it is applied to each sample using a specialized piece of equipment, an alternating field demagnetizer modified so that a small, constant, biasing magnetic field can be applied during demagnetization. After application of the ARM, the sample's magnetic intensity is measured in a rock magnetometer. Usually a superconducting rock magnetometer is used for speed of measurement and high accuracy. The measurement of each sample takes on the order of about 5–7 minutes for both the ARM application and the magnetic intensity measurement. Five hundred samples can be processed in about 50–60 hours, once they have been trimmed and weighed.

Time series analysis (Chapter 4) is used to deconstruct the ARM data series into its constituent frequencies. Determining whether any cycles observed are Milankovitch (astronomically forced) is the most difficult part of the study. Ideally, some independent control on time is available to unequivocally identify astronomically forced cycles. Biostratigraphy or magnetostratigraphy (Chapter 3) are two important ways of doing this. The time assignment can be at a fairly coarse scale, just high enough resolution to allow the identification of the longest (~100 or 405 kyr eccentricity) astronomically forced cycles targeted for detection in the sampling. Once these cycles are identified, the series can be tuned either to a theoretical insolation series (e.g., Laskar et al. 2004) or to a simple sinusoid at that frequency to remove the effects of varying sedimentation rates in the record. Time series analysis of the tuned series can determine whether the shorter astronomically forced cycles (precession, obliquity) are more pronounced in the power spectrum, thus providing further evidence that astronomically forced cycles have been identified. The identification of astronomical cycles has now translated the coarse resolution biostratigraphy or magnetostratigraphy to a high-resolution chronostratigraphy for the sedimentary sequence.

1.3 The Significance of Rock Magnetic Cyclostratigraphy

The realization of a high-resolution chronostratigraphy using rock magnetic cyclostratigraphy has the potential to be a transformative chronostratigraphic technique for the Earth sciences. High-resolution time can be critical to understanding many important Earth processes. Rock magnetic cyclostratigraphies have already been used to study tectonic processes. Gunderson et al. (2012) have used a susceptibility record of obliquity in Plio-Pleistocene marine sediments from the Po River Valley in northern Italy to time the deposition of growth strata that, in turn, constrain the folding of the Salsomaggiore anticline, and hence fault slip on the blind thrust pushing up the folded rocks. A similar study of folded Eocene marine growth strata in the Spanish Pyrenees is the ultimate goal for the ARM cyclostratigraphy already produced for the Arguis Formation marine marls (Kodama et al. 2010). Not only high-resolution chronostratigraphies will result from rock magnetic cyclostratigraphies, but high-resolution correlation of sedimentary sections that are distant globally, or even cores drilled from an oil or gas field for petroleum exploration and exploitation.

Rock magnetic cyclostratigraphy also has the potential for outperforming the resolution of radioisotope geochronology, particularly for very ancient rocks. Minguez et al. (2014) have demonstrated rock magnetic cyclostratigraphies at different localities of the Neoproterozoic Johnnie Formation from near Death Valley, CA. These chronostratigraphies yield precession-scale resolution that in the Neoproterozoic is on the order of 15 kyr. Bowring and Schmitz (2003) indicate errors of approximately ±0.3–3 Ma on Neoproterozoic age (~550 Ma) rocks in their discussion of the problems dating the Cambrian–Neoproterozoic boundary with zircon U–Pb ages. Even the 555.0 ± 0.3 Ma age they report for the Neoproterozoic has an error 20 times larger than the precision realized by the Johnnie Formation rock magnetic cyclostratigraphy. Furthermore, radioisotopic dates give spot ages at irregular intervals, whereas cyclostratigraphy yields a continuous chronology and preserves high-precision ages through several million year intervals.

1.4 Layout of the Book

This book is organized to provide the background information needed to conduct a rock magnetic cyclostratigraphic study of a sedimentary sequence. Chapter 2 covers the important points about rock magnetics necessary for conducting rock magnetic cyclostratigraphy studies and for understanding the interpretation of rock magnetic data. Chapter 3 is a primer for the basics of conducting a magnetostratigraphic study needed to assign time at a coarse scale to the data series acquired in a cyclostratigraphic study. Chapter 4 covers the basics of time series analysis needed to extract and interpret cycles in rock magnetic data series; step-by-step procedures and commands are demonstrated with MATLAB® scripts. Chapter 5 gives the theoretical background of astronomical forcing mechanisms and provides step-by-step description for how to calculate obliquity and precession index solutions using FORTRAN code. Chapter 6 presents case studies of rock magnetic cyclostratigraphy. The Appendix contains scripts and codes used throughout the book.

References

Bowring, S.A. & Schmitz, M.D. (2003) High precision U-Pb zircon geochronology and the stratigraphic record. *Reviews in Mineralogy and Geochemistry, 53*, 305–326. DOI:10.2113/0530305.

deMenocal, P.B. & Bloemendal, J. (1995) Plio-Pleistocene climatic variability in subtropical Africa and the palaeoenvironment of hominid evolution: A combined data-model approach. In: Vrba, E.S., Denton, G.H., Partridge, T.C., & Burckle, L.H. (eds), *Paleoclimate and Evolution, with Emphasis on Human Origins*, pp. 262–288. Yale University Press, New Haven.

Evans, M.E. & Heller, F. (2003) *Environmental Magnetism: Principles and Applications of Enviromagnetics*, 299 pp. Academic Press, Amsterdam.

Gunderson, K.L., Kodama, K.P., Anastasio, D.J., & Pazzaglia, F.J. (2012) Rock-magnetic cyclostratigraphy for the Late Pliocene-Early Pleistocene Stirone section, Northern Apennnine mountain front, Italy. *Geological Society, London, Special Publications, 373*, 26. DOI:10.1144/SP373.8.

Hays, J.D., Imbrie, J., & Shackleton, N.J. (1976) Variations in the Earth's orbit: Pacemaker of the ice ages. *Science, 194*, 1121–1132. DOI:10.1126/science.194.4270.1121.

Heller, F., Forster, T., Evans, M.E., Bloemendal, J., & Thouveny, N. (1998) Gesteinsmagnetische archive globaler Umweltanderung. *GeoArchaeoRhein, 2*, 151–162.

Kissel, C., Laj, C., Labeyrie, L., Dokken, T., Voelker, A., & Blamart, D. (1999) Rapid climate variations during marine isotopic stage 3: Magnetic analysis of sediments-from Nordic Seas and North Atlantic. *Earth and Planetary Science Letters, 171*, 489–502. DOI:10.1016/S0012-821X(99)00162-4.

Kodama, K.P., Anastasio, D.J., Newton, M.L., Pares, J., & Hinnov, L.A. (2010) High-resolution rock magnetic cyclostratigraphy in an Eocene flysch, Spanish Pyrenees. *Geochemistry, Geophysics, Geosystems, 11*. DOI:10.1029/2010GC003069.

Laskar, J., Robutel, P., Joutel, F., Gastineau, M., Correia, A.C.M., & Levrard, B. (2004) A long term numerical solution for the insolation quantitties of the Earth. *Astronomy & Astrophysics, 428*, 261–285. DOI:10.1051/0004-6361:20041335.

Minguez, D.A., Kodama, K.P., & Hillhouse, J.W. (2014) Paleomagnetic and cyclo-stratigraphic constraints on the duration of the Shuram carbon isotope excursion, Johnnie Formation, Death Valley region, CA, Geochem., Geophys., Geosys. (in review).

Olsen, P.E. & Kent, D.V. (1996) Milankovitch climate forcing in the tropics of Pangea during the Late Triassic. *Palaeogeography, Palaeoclimatology, Palaeoecology, 122*, 1–26. DOI:10.1016/0031-0182(95)00171-9.

Sadler, P.M. (1981) Sedimentation rates and the completeness of stratigraphic sections. *Journal of Geology, 89*, 569–584. DOI:10.1086/628623.

Sagnotti, L., Florindo, F., Verosub, K.L., Wilson, G.S., & Roberts, A.P. (1998) Environmental magnetic record of Antarctic palaeoclimate from Eocene/Oligocene glaciomarine sediments, Victoria Land Basin. *Geophysical Journal International, 134*, 653–662. DOI:10.1046/j.1365-246x.1998.00559.x.

Stoner, J.S., Channell, J.E.T., & Hillaire-Marcel, C. (1995) Magnetic properties of deep-sea sediments off southwest Greenland: Evidence for major differences between the last two deglaciations. *Geology*, *23*, 241–244. DOI:10.1130/0091-7613(1995)023<0241:MPODSS>2.3.CO;2.

Weedon, G.P. (2003) *Time-Series Analysis and Cyclostratigraphy: Examing Stratigraphic Records of Environmental Cycles*, 259 pp. Cambridge University Press, Cambridge.

2

Rock Magnetism

Abstract: This chapter covers the basics of rock magnetism needed for conducting a rock magnetic cyclostratigraphy study. The fundamental types of material magnetic behavior, **diamagnetism, paramagnetism, and ferromagnetism,** and their causes are discussed. The basic characteristics and properties of the important ferromagnetic minerals—the iron oxides, iron oxyhydroxides, and iron sulfides—are summarized. The fine particle magnetic behavior that governs the magnetic particles carrying a rock magnetic cyclostratigraphy is introduced. Fine particle magnetism topics that are discussed include hysteresis, individual magnetic particle anisotropy, and magnetic domains. Finally, the environmental magnetic parameters, and their ratios, that are important to rock magnetic cyclostratigraphy are explained. The chapter ends by discussing the importance of identifying the magnetic mineralogy in a sedimentary sequence targeted for a rock magnetic cyclostratigraphy study and choosing the best environmental magnetic parameter for the cyclostratigraphic study.

2.1 Introduction

The basic idea of rock magnetic cyclostratigraphy is to measure the variations in one or several rock magnetic parameters throughout a sequence of sedimentary rocks. The rock magnetic parameter is selected based on its ability to be a proxy for an environmental or climate process at the time of the sediment's deposition. Rock magnetic cyclostratigraphy uses the principles of environmental magnetism, an important subdiscipline of rock magnetism (Thompson & Oldfield 1986; Evans & Heller 2003; Tauxe 2010; Kodama 2012), in which magnetic measurements of a sample can detect one of the three important magnetic characteristics: the concentration of a particular magnetic mineral, the grain size of the magnetic mineral particles, and the different types of magnetic minerals in a sample. These measurements, which will be covered in more detail later in this chapter, are typically done by the application of

Rock Magnetic Cyclostratigraphy, First Edition. Kenneth P. Kodama and Linda A. Hinnov.
© 2015 John Wiley & Sons, Ltd. Published 2015 by John Wiley & Sons, Ltd.

remanent magnetizations to samples in the laboratory and then, of course, the measurement of the resulting magnetization. The use of environmental magnetism to study either recent or past changes in the environment or climate has reached fairly sophisticated levels, but rock magnetic cyclostratigraphy has kept to simpler applications of environmental magnetism by measuring magnetic mineral concentration variations to construct a data sequence for time series analysis.

In contrast, environmental magnetic studies can examine many different environmental processes including modern soil development, particularly the magnetic enhancement and growth of new magnetic minerals in the topsoil, paleosol development in loess by the growth of secondary magnetic minerals and their connection to paleo-precipitation, reductive diagenesis in marine and lacustrine, organic-rich sediments, erosion and transport of catchment soils and bedrock into either a lake or river, ancient lake level fluctuations, population fluctuations of magnetotactic bacteria in lake or marine sediments, and the relative contributions of riverine and eolian sources for near-shore marine sediments. There are many more examples that can be found in excellent summaries of the environmental magnetic literature (Thompson & Oldfield 1986; Maher et al. 1999; Evans & Heller 2003; Liu et al. 2012).

One consideration is of utmost importance for any rock magnetic cyclostratigraphic study. It is absolutely critical that the magnetic minerals in the sedimentary rocks being studied are primary depositional minerals indicating that their concentration or grain size variations reflect depositional processes. Therefore, before a comprehensive cyclostratigraphic study is conducted, the age of the magnetic minerals should be studied. The best way to determine the age of the magnetic minerals in a sedimentary rock is to conduct a pilot study of the paleomagnetism carried by the magnetic minerals. The age of the resulting paleomagnetic directions can be constrained by standard paleomagnetic tests. The fold test determines whether the paleomagnetism of a sequence of folded sedimentary rocks is older or younger than the time of folding. If the magnetization fails the fold test and is younger than folding, the magnetic minerals are clearly secondary and undoubtedly unsuited for a rock magnetic cyclostratigraphic study. However, even if the paleomagnetism passes the fold test and is older than the folding, further tests should be conducted to constrain the age of the magnetic minerals, since the folding may not have occurred soon enough after deposition to ensure that the magnetic minerals are depositional in age. Passage of the fold test shows, at least, that the magnetic minerals are ancient and older than the folding. Other tests constraining the age of magnetization, and the magnetic minerals carrying the magnetization, are the baked contact test and the conglomerate test. These tests and the fold test are described in more detail in standard paleomagnetism texts. Butler (1992) and Tauxe (2010) provide excellent summaries of these tests.

One powerful demonstration of primary magnetic minerals in a sedimentary sequence is the successful determination of a magnetostratigraphy, particularly if the magnetostratigraphy can be easily correlated to the Geomagnetic polarity time scale (Gradstein et al. 2004; Gradstein et al. 2012). Since the measurement of a magnetostratigraphy is an important way of assigning time to a stratigraphic section so that Milankovitch cycles are more easily identified, in many rock magnetic cyclostratigraphic studies, a magnetostratigraphy has been determined as a part of the study, and in doing so has demonstrated that the magnetic minerals are likely to be primary, depositional minerals. In some cases, a previous paleomagnetic or magnetostratigraphic study has shown that the paleomagnetism is depositional in age and that the rocks are perfect for a rock magnetic cyclostratigraphic study. In some cases, poor paleomagnetic results do not necessarily rule out depositional magnetic minerals. For instance, Precambrian and Paleozoic rocks of the southern California–western Nevada region are dominated by secondary viscous magnetizations (Gillett & Van Alstine 1982) that are carried by depositional minerals. The rocks have a poor magnetostratigraphy but have yielded a reasonable rock magnetic cyclostratigraphy (see Chapter 6; Minguez et al. 2014).

In the rock magnetic cyclostratigraphies described in this book, the emphasis is on the permanent or remanent magnetizations of the rocks. Successful cyclostratigraphies have resulted from the measurement of magnetic susceptibility variations throughout a sedimentary sequence (see, for example, Mayer & Appel 1999; Ellwood et al. 2011; Gunderson et al. 2012); however, as will be shown in Section 2.4, the remanent magnetization measurements typically used for rock magnetic cyclostratigraphy can be targeted for specific magnetic minerals or subpopulations of a magnetic mineral. The induced magnetization that is measured in a susceptibility measurement is the composite of the induced magnetizations for the different magnetic minerals in a sample: iron-rich silicates or clays, calcite or quartz grains, and the ferromagnetic mineral grains of either very small or large grain size. The susceptibility signal cannot be easily deconstructed, so the source of the cyclostratigraphic signal is usually poorly known and the mechanism by which the environmental or climate process was encoded remains somewhat of a mystery.

2.2 Types of Magnetism

Understanding the concept of a magnetic dipole moment is critical to understanding magnetization measurements for rock magnetic cyclostratigraphy. The magnetization or magnetism of any material can be envisioned as a dipole moment. A dipole moment has two magnetic poles, a north (+) pole and a south (–) pole that are always paired. The magnetic moment vector points from the south pole toward the north pole (Figure 2.1). Unlike an electric dipole moment, in which positive and negative electric charges become paired,

(a)

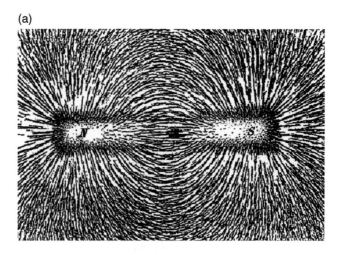

(b)

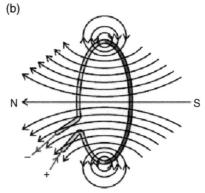

Figure 2.1 Field lines of a magnetic dipole field. (a) The classic experiment of iron filings following the field lines of a bar magnet (This figure is in the public domain in the United States). (b) Dipole field lines created by an electric current flowing in a loop. Source: http://librarykvbirbhum.wordpress.com/physics-forums/.

but can be split apart into separate electric charges (e.g., electrons and positrons), the magnetic poles in a magnetic dipole cannot be separated into individual magnetic monopoles. This observation makes Maxwell's equations governing electricity and magnetism asymmetric and has been a source of puzzlement to physicists, but the practical meaning is that no matter how much you subdivide magnetic material, it only breaks into smaller and smaller magnetic dipole moments. It is the fundamental nature of material magnetism. A magnetic dipole moment can also be generated by an electric current (moving electric charges) flowing in a loop (Figure 2.1).

A magnetic dipole moment generates a magnetic field with a characteristic geometry (Figure 2.2) in which field lines emerge from the N (+) magnetic pole and loop around to enter into the S (−) magnetic pole. The magnetic dipole field can be described by the following equations:

$$H_\theta = \frac{M\sin\theta}{r^3} \tag{2.1}$$

$$H_R = \frac{2M\cos\theta}{r^3} \tag{2.2}$$

Figure 2.2 (a) geometry for calculating the dipole field for the dipole moment shown by the arrow using Equations (2.1) and (2.2) in the text. P is the point where the field is calculated. (b) Interference pattern for the radial fields for two monopoles paired together to create a dipole. The dipole field is the result of the interference between two radial fields for two sources close together, viewed from a distance (the far field).

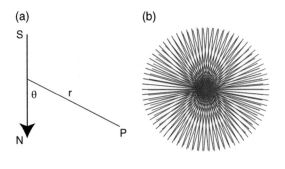

where θ is the angle between the dipole moment and the radial vector out to the point in space where the field is calculated, r is the distance from the dipole moment out to the point in space, and M is the dipole moment's strength (Figure 2.2). The equations above are written for cgs (centimeter–gram–second) units; for SI (System International) units, the denominators would also include a 4π factor multiplying the cubed distance term. One way of conceptualizing the dipole moment is that it is the result of the interference between the radial field lines of separate (and imagined) magnetic monopoles when observed in the far field, i.e., the monopoles are very close together compared to the distance of the observation point from the dipole (Figure 2.2). The magnetic measurements made in a paleomagnetism laboratory or rock magnetism laboratory are in reality the measurements of the dipole fields generated by the samples. Typically for rock magnetic cyclostratigraphy measurements, the material's magnetic dipole moment is normalized by the mass of the material. Converting from cgs magnetics units to SI units is tricky and often involves a 4π factor, but the conversion from mass normalized cgs units (emu/g) to mass normalized SI units (Am²/kg) is easy; the conversion factor is simply unity. It is very important to keep track of units in any rock magnetic study, since proper units are critical to quantitative comparison of results between different environmental magnetic studies. Tauxe (2010) and Butler (1992) both provide good discussions of magnetic units and the conversion between cgs units, which are typically used in most paleomagnetic laboratories, and SI units, which must be used for publication. However, for convenience, Table 2.1 gives the critical conversions between cgs and SI magnetic units for magnetization, field, and susceptibility.

All materials, either natural or human-made, are magnetic in that either they carry a permanent magnetization (remanent magnetization) or they become magnetized during exposure to a magnetic field (induced magnetization). Basically, there are three different types of material magnetism: diamagnetism, paramagnetism, and ferromagnetism.

Diamagnetism is a weak induced magnetization in which a material acquires a magnetization with a magnetic moment opposite in direction to

Table 2.1 Conversion between SI and cgs magnetic units

Magnetic Parameter	SI unit	cgs unit	Conversion
Magnetic moment	Amp-m^2 (Am^2)	emu	1 Am^2 = 10^3 emu
Magnetization/mass	Am^2/kg	emu/g	1 Am^2/kg = 1 emu/g
Magnetic field (H)	A/m	Oersted (Oe)	1 A/m = 4π × 10^{-3} Oe
Magnetic induction (B)	Tesla (T)	Gauss (G)	1 T = 10^4 G
Susceptibility/mass	m^3/kg	emu/g/Oe	1 m^3/kg = (10^3/4π) cgs

Source: B = μ(H + M) where M is the magnetization of a material and μ is the magnetic permeability.

the applied magnetic field. All materials are diamagnetic, but in some materials stronger magnetizations (e.g., paramagnetism and ferromagnetism) swamp the diamagnetism. Diamagnetism is due to the interaction of the magnetic moment of orbiting electrons with the applied magnetic field. Orbiting electrons are like current loops and generate dipole moments. Calcite and quartz are important rock-forming minerals that are diamagnetic. Water and organic material are also diamagnetic. Typical magnitudes of diamagnetic susceptibilities are as follows: quartz = -0.62×10^{-8} m^3/kg, calcite = -0.48×10^{-8} m^3/kg, and water = -0.9×10^{-8} m^3/kg.

Paramagnetism is an induced magnetization in which a material acquires a magnetization with a magnetic moment parallel to the direction of the applied magnetic field. Paramagnetism is due to materials containing atoms with net magnetic moments because of unpaired electrons. Electrons have spin and the spin of an electric charge generates a magnetic dipole moment. Paired electrons means that two electrons occupy the same atomic orbital but their spins are opposite to each other. Therefore, their spin magnetic moments cancel out. The unpaired electron spin moments, resulting when only one electron occupies an orbital, will align with an applied magnetic field. Iron has unpaired electrons in its 3D electron shell, so iron-containing silicates (e.g., amphiboles, pyroxenes, and iron-rich clays) are paramagnetic. Paramagnetism is about 100 times stronger than diamagnetism. Values for common Fe-rich Earth materials are as follows: biotite = 79×10^{-8} m^3/kg and nontronite (Fe-rich clay) = 65×10^{-8} m^3/kg. Paramagnetism is temperature dependent and this dependence is described by the Curie law of paramagnetism (see Butler (1992) for complete coverage) which indicates that the strength of the induced paramagnetism is dependent on the following:

$$J_{ind} \propto \frac{MH}{kT} \qquad (2.3)$$

where J_{ind} is the induced paramagnetism, M is the moment resulting from the unpaired electrons, and H is the applied field that rotates the moment into alignment. T is the temperature multiplied by Boltzmann's constant, k, in the denominator. The denominator represents the misaligning influence of thermal energy.

The third type of material magnetism is the kind most important to rock magnetic cyclostratigraphy, ferromagnetism. Ferromagnetic substances are permanently magnetized; they remain magnetized even in the absence of an applied magnetic field. Remanence is the permanent magnetization of ferromagnetic materials. The apocryphal story is that the first description of the remnant of magnetization left over when the applied field was turned off was misspelled as remanent. Ferromagnetism results from a cooperative interaction between the unpaired spin moments of electrons in adjacent atoms called exchange interactions. Only three elements can form ferromagnetic substances: iron, nickel, and cobalt. Iron is the fourth most common element in the Earth's crust; so ferromagnetic materials, such as iron oxides, iron sulfides, and iron oxyhydroxides, are almost ubiquitous. In iron, the unpaired 3D electrons have their spin moments aligned by exchange interactions giving the crystal lattice a magnetic order and hence a permanent magnetization. Strictly speaking, when the adjacent atom's spin moments are aligned parallel, it is called ferromagnetism, and when the coupling is antiparallel, it is called anti-ferromagnetism. A third subset of ferromagnetism is called ferrimagnetism and is the manner by which magnetite, a very important magnetic mineral, acquires its magnetization. In ferrimagnetism, sublattices of the crystal have spin moments that are aligned opposite to each other, but there is a different number of spin moments in each sublattice, so a net magnetization results.

Rock magnetism, and its application to environmental magnetism, is the study of the remanent magnetization of fine particles (micron to submicron in size). These small particles carry the paleomagnetism recording the ancient geomagnetic field and retain records of environmental processes during deposition.

2.3   Ferromagnetic Minerals

Iron is the fourth most common element in the Earth's crust, after oxygen, silicon, and aluminum, and makes up about 5.6% of crustal rocks; therefore, tracing the movement of iron, by magnetic methods, can be a powerful way of tracing the movement of Earth materials between the reservoirs of the Earth system, as envisioned in Earth system science (Thompson & Oldfield 1986; Liu et al. 2012). Natural magnetic minerals, that always contain iron, move between the lithosphere, the hydrosphere, the atmosphere, and the biosphere. Iron is not usually found in its native state in the Earth because of the high oxygen content of the atmosphere, and the most important and most common magnetic (i.e., ferromagnetic) minerals are the iron oxides, magnetite (Fe_3O_4), and hematite (Fe_2O_3). Ferromagnetic materials have orders of magnitude stronger susceptibilities than diamagnets and paramagnets ($1,000–10,000 \times 10^{-8}$ m^3/kg), so a very small concentration of ferromagnetic minerals will swamp the induced magnetizations of diamagnetic and paramagnetic minerals in a sample.

Magnetite is ferrimagnetic and hence has a strong spontaneous magnetization (92 Am^2/kg) resulting from exchange interactions between the unpaired spin moments of its 3D electrons. It has cubic crystal symmetry and is an inverse spinel. Butler (1992) and Tauxe (2010) provide more detailed information about the crystal structure of magnetite and the origin of its spontaneous magnetization. Magnetite can be formed in igneous rocks and it is magnetized as it cools down through its Curie temperature (580°C) and the ferromagnetic exchange interactions are established. Titanium can fit into magnetite's crystal structure and a solid solution series exists between antiferromagnetic ulvospinel (Fe$_2$TiO$_4$) and magnetite. As more Ti is added to the crystal structure, the Curie temperature drops to low values of about –100°C for ulvospinel. An important titanomagnetite is TM60 that is 60% ulvospinel and is the ferromagnetic mineral that crystallizes in mid-ocean ridge basalts that make up the seafloor. Although magnetite is typically considered to be a primary depositional magnetic mineral, it can form long after deposition as a secondary magnetic mineral, by tectonically driven fluid flow (McCabe & Elmore 1989) or burial diagenesis of clays (Woods et al. 2002). See Kodama (2012) or Van der Voo and Torsvik (2012) for a summary of remagnetization processes. The grain size of magnetite particles is important in paleomagnetic and environmental magnetic studies. The magnetite particles important for carrying a stable paleomagnetic signal are micron to submicron in size, and the principles of fine particle magnetism describe their behavior. Magnetite has iron atoms in both the 2+ and 3+ oxidation states, so magnetite can be oxidized either to maghemite or hematite.

In rock magnetic cyclostratigraphy, it is important to know what magnetic minerals are carrying cyclic behavior, particularly if it is identified as orbitally forced cyclicity. Also, in designing the cyclostratigraphic study, it will be important to know what laboratory-applied remanence will best measure the concentration variations or grain size variations in the sedimentary sequence. Magnetic minerals can be identified with two important pieces of information, their coercivity or magnetic hardness, which will be explained in Section 2.4 and their Curie temperature. Low temperature (<< room temperature) behavior can also be diagnostic, but specialized equipment is needed to make low temperature measurements. Table 2.2 summarizes some diagnostic characteristics for the main magnetic minerals that are the target of rock magnetic cyclostratigraphic studies, as well as the spontaneous magnetization for these minerals.

Hematite is the second important ferromagnetic mineral (αFe$_2$O$_3$). Hematite has hexagonal crystal symmetry with a canted anti-ferromagnetism. Sublattices of its crystal have their 3D spin moments aligned antiparallel to each other, but they are canted slightly from being exactly antiparallel, by <0.1°, so a net, but

Table 2.2 Important rock magnetic properties.

Magnetic mineral	Spontaneous magnetization (Am²/kg)	Range of coercivity	Curie temperature (°C)
Magnetite	92	10–100 mT	580
TM60	24	~8 mT	150
Hematite	0.4	100s of mT to several T	680
Greigite	25	60 > 100 mT	~320
Goethite	Varies ≲1	5–10 T	~125
Maghemite	74	10–100 mT	Inverts to hematite at about 300–350

Source: Table based on the information in Tauxe (2010), O'Reilly (1984), Dunlop and Ozdemir (1997).

weak, ferromagnetism results (spontaneous magnetization = 0.4 Am²/kg). Hematite can be a primary magnetic mineral, but it is more likely to be a secondary magnetic mineral. Many times it can be formed early in the postdepositional history of a sediment by secondary chemical growth, in which case it is magnetized by its growth in the Earth's magnetic field and acquires a **chemical or crystallization remanent magnetization (CRM)**. Secondary hematite is formed by the oxidation of Fe-rich silicates probably in arid environments or perhaps during the dry season of monsoonal climates (Kodama 2012). Many red bed sedimentary sequences, whose red color is due to very fine-grained (submicron), pigmentary hematite and whose paleomagnetic signal is carried by either the pigmentary hematite or large grain size (micron size) specular hematite particles, can record a magnetostratigraphy and show evidence of inclination shallowing. Inclination shallowing, in which the angle that the paleomagnetic vector makes with the paleohorizontal is smaller than the geomagnetic field in which the rocks were deposited, suggests that the hematite is either depositional or formed very soon after deposition, so that burial compaction will affect its inclination (Kodama 2012). A magnetostratigraphy carried by hematite is also evidence of either a primary depositional remanence or a CRM acquired soon after deposition. The pigmentary hematite in red bed sequences is usually secondary and probably formed on the order of 10^5–10^6 years after deposition (Kodama 2012).

Iron sulfides are formed during the reductive diagenesis that occurs in organic-rich sediments, both in marine and lacustrine settings. Iron sulfides are clearly secondary but can be formed soon after deposition (10^3–10^5 years) in the top meter of the sediment column (see Table 6.1 in Kodama 2012). During reductive diagenesis, the primary, depositional iron oxide (typically magnetite) is dissolved and then a sequence of iron sulfides are formed, including greigite (Fe_3S_4), as intermediate products, with pyrite as the final mineral in the sequence. Pyrite is paramagnetic, but ferrimagnetic greigite is strongly magnetic and can carry a stable paleomagnetic signal. If the sediment accumulation rate is fast enough, sediments pass through the reductive diagenesis zone quickly enough so the reaction doesn't go to completion, and greigite, rather than pyrite, is left in the sediments. Also, it is likely that the magnetite may not be completely dissolved, so the sediments are left with a mixture of primary depositional magnetite and secondary greigite. Since magnetite and greigite have similar coercivities; they can both contribute to a rock magnetic cyclostratigraphy if the low coercivity magnetic minerals are activated to measure concentration variations. The only way to separate their contributions to the laboratory remanence applied for a rock magnetic cyclostratigraphy is by thermal demagnetization since greigite has a Curie temperature of about 300°C. However, heating organic-rich sediments often leads to oxidation and formation of secondary magnetite. Pyrrhotite (Fe_7S_8) is another important ferromagnetic iron sulfide mineral. It has been viewed as a product of reductive diagenesis, but Horng and Roberts (2006) provide evidence that it forms very slowly at temperatures lower than 180°C and that greigite is the only important magnetic mineral formed by reductive diagenesis.

Goethite (FeOOH) is an anti-ferromagnetic mineral that can have a spontaneous magnetization due to lattice vacancies. It is formed by oxidation of Fe-rich precursor minerals such as Fe-rich clays or other Fe-rich silicates. It has a very low Neel temperature at which it loses its spontaneous magnetization (~125°C), but extremely high coercivities. Both properties make it easy to identify. Most paleomagnetic evidence suggests that goethite often forms by recent weathering, since it often carries a paleomagnetic direction parallel to the present day geomagnetic field. However, there is evidence that the ratio of ancient goethite to hematite in a rock is a sensitive indicator of moisture in the sediment's source area (Yapp 2001; Harris & Mix 2002). For this reason, magnetic measurement of the goethite/hematite ratio could be an important paleoclimate indicator for rock magnetic cyclostratigraphy; however, it will be important to check, using paleomagnetism, whether the goethite is ancient or a product of present day weathering. Maghemite (γFe_2O_3) is another important secondary magnetic mineral. It is the product of low-temperature oxidation of magnetite. It retains the crystal structure of magnetite but has the chemical formula of hematite. Lattice vacancies are left in the crystal as about one-third of the Fe^{2+} ions migrate to the surface for oxidation, so its magnetic properties can vary, like those of goethite. The remaining Fe^{2+} cations are oxidized but remain in the crystal lattice. Maghemite is often found to form in modern soils, making it potentially a depositional magnetic mineral for sedimentary rocks derived from soil. A good way to identify maghemite is by heating. Although its Neel temperature is high (~645°C), it inverts to hematite at about 350°C.

2.4   Fine Particle Magnetism

2.4.1   *Hysteresis*

Ferromagnetism can be best described with a hysteresis plot that shows the magnetic behavior of a collection of small ferromagnetic particles while a field is being applied (Figure 2.3). There are four important parameters that can be garnered from a hysteresis plot that describe the characteristics of the ferromagnetic grains in the sample: the coercivity (H_c or B_c), the coercivity of remanence (indicated by either H_{cr} or B_{cr}), the **saturation magnetization** (J_{sat} or M_s), and the saturation remanence (either J_{rs} or M_{rs}). This confusing array of letters comes from the fact that H represents magnetic field and B represents magnetic induction. B is the combination of the magnetic field and the magnetization induced in matter (Table 2.1). Since the magnetization induced in air is essentially zero, B and H are virtually have the same value in air, where most hysteresis measurements are made, and B and H are used interchangeably. M stands for magnetic moment, and J for magnetization, or magnetic moment normalized by either mass or volume, so they both represent the magnetism induced in the sample by application of a magnetic field. For the purposes of this book, we'll stick to B and J, but the

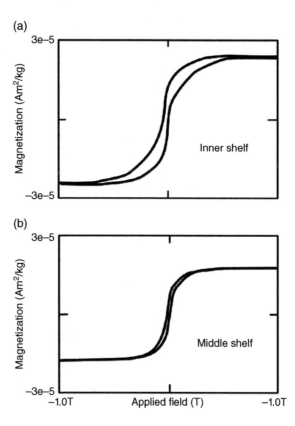

Figure 2.3 Example hysteresis loops for samples from the Cretaceous Cupido Formation from Mexico (see Chapter 6). The constricted "waist" in the top loop probably comes from a mixture of different size ferromagnetic grains. Source: Hinnov, Kodama, Anastasio, Elrick & Latta 2013. Copyright 2013 by the Geological Society of London.

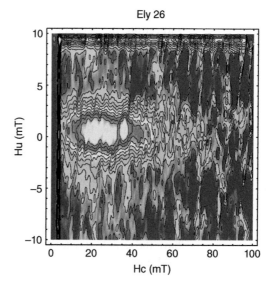

Figure 2.4 FORC distribution diagram from a Lake Ely sediment sample showing a 10–40 mT coercivity range for the ferromagnetic magnetofossils in the sediment. The spread on the H_u axis suggests some magnetic interactions between the particles.

reader is warned that M and H, meaning essentially the same parameters, can show up in different publications. The ratios of the four hysteresis parameters indicate something about the domain state (see Section 2.4.3) of the magnetic grains and hence something about their magnetic grain size. A more sophisticated use of hysteresis behavior is **first-order reversal curve (FORC)** distribution diagrams (Figure 2.4), in which multiple hysteresis loops are acquired for a sample over a great range of field strengths. Analysis of FORC diagrams indicates not only domain state and hence magnetic particle grain size but also whether there are magnetic interactions between the magnetic grains in a sample. Tauxe (2010) provides a more detailed explanation of FORC diagrams.

The main point about the hysteresis loop is it indicates that ferromagnetic particles have irreversible magnetic behavior. As the applied field (B) is increased, the magnetization (J) of the sample increases in the same direction. If the field is only increased to small values, similar to that of the Earth's field (~50 μT), when the field is turned off, the magnetization of the sample also returns to zero. This is the only reversible behavior for a ferromagnetic substance and is denoted as ferromagnetic susceptibility. It is much greater in magnitude than paramagnetic susceptibility, but the macroscopic behavior is similar. If the field is increased beyond these low values, the magnetic behavior is no longer reversible and when the field is turned off, the sample retains a magnetization called an **isothermal remanent magnetization (IRM)**. If the field is then increased and the magnetization levels out and no longer increases, the magnetization is saturated and denoted by J_{sat}. J_{sat} is a direct measure of the total number of magnetic grains in a sample, but can only be measured while the field is applied. In most paleomagnetic laboratories the magnetometers used can only measure the sample when the field is turned off (and, in fact, brought as close to zero as possible). In this case, if the sample has reached saturation, the saturation remanence is obtained (J_{rs}) when the field is reduced to zero. For a collection of small, **single domain (SD)** grains, the ratio of J_{rs} to J_{sat} is 0.5, if the magnetizations of the grains are randomly oriented in space. J_{rs} is also called the **saturation isothermal remanent magnetization (SIRM)**.

In order to decrease a sample's magnetization to zero, i.e., to totally demagnetize a sample, the field must be applied in the opposite direction. When the sample's magnetization is zero, with the field still turned on, the value of the field needed to maintain zero magnetization is called the coercivity of the sample (B_c). The coercivity is a measure of the "hardness" of the sample's magnetization, or how much the sample needs to be coerced with an applied field to force its magnetization down to zero. If, however, the field is then turned off, the magnetization will return to some positive value. In order to have the sample stay totally demagnetized when the field is turned off, the field must be increased in the opposite direction to the coercivity of remanence value (B_{cr}) that is greater than the coercivity (B_c). The sample actually becomes magnetized in the opposite direction while the field is applied, but when the field is turned off, the sample's magnetization goes to zero and is completely demagnetized.

Often the hysteresis parameters are plotted as ratios and are used to easily summarize the magnetic characteristics of the sample with one point on a plot. This is called the Day plot (Day et al. 1977) in which J_{rs}/J_{sat} is plotted as function of B_{cr}/B_c (Figure 2.5). Points in the upper left hand corner of the plot ($J_{rs}/J_{sat} \sim 0.5$ and $B_{cr}/B_c \sim 0.5$) indicate SD magnetic behavior while points in the lower right ($J_{rs}/J_{sat} \sim 0.05$, $B_{cr}/B_c \sim >5$) indicate a collection of grains with many magnetic domains. Unfortunately, in reality, many samples plot in the central region of the plot indicating either a third kind of magnetic behavior: **pseudo single domain (PSD)** (grains that have multiple domains but behave like they have just one domain) or a mixture of SD and **multidomain (MD)** grains. Because the Day plot isn't too useful in distinguishing magnetically between samples, it has fallen out of favor for describing the magnetization of samples. A modification of the Day plot in which J_{rs}/J_{sat} is plotted as a function of B_c is also used to evaluate magnetic particle size (Figure 2.5). FORC diagrams are currently considered one of the best ways to describe a sample's ferromagnetic behavior.

Before starting a rock magnetic cyclostratigraphic study, it is important to identify the magnetic minerals in the rocks so the appropriate rock magnetic parameter can be chosen to construct the time series. A critical determination is whether the magnetic mineralogy of the rocks is dominated by magnetite or hematite. One of the best ways to make the determination is to conduct an IRM acquisition experiment. In this experiment, a sample is exposed to higher and higher DC magnetic fields, usually in an impulse magnetizer that creates a brief, but strong, magnetic field by discharging a capacitor through a coil that surrounds the sample. After exposure to the field, the sample is measured. The field is increased until the

(a)

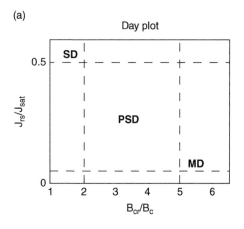

(b)

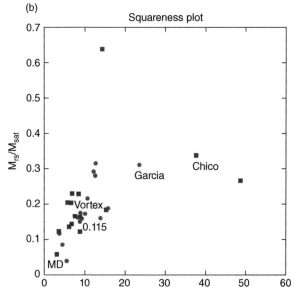

Figure 2.5 Day plot of hysteresis parameter ratios: J_{rs}/J_{sat} plotted as a function of B_{cr}/B_c (a). "Squareness plot" of M_{rs}/M_{sat} vs. H_c for samples from the Cupido Formation of Mexico (b). This type of plot is sometimes used instead of the Day plot to indicate ferromagnetic particle domain state and hence grain size. MD, multidomain; vortex is a PSD state of Tauxe et al. (2002). Source: Hinnov, Kodama, Anastasio, Elrick & Latta 2013. Copyright 2013 by the Geological Society of London.

remanent magnetization reaches saturation and no longer increases, the SIRM or J_{rs} has been reached. There may be different magnetic minerals and populations of grains with different magnetic mineral grain sizes in the sample. Each of these magnetic minerals or population of magnetic grains will have a different mean coercivity of remanence (B_{cr}) which can be modeled by fitting Gaussian curves to the first derivative of the IRM acquisition curve (Kruiver et al. 2001) (Figure 2.6). Magnetite has much lower coercivities than hematite (Table 2.2) and the coercivity analysis of the IRM acquisition curve can be used to distinguish between magnetite and hematite in most cases. If there is the possibility that the iron sulfide greigite is present in the samples, it cannot be easily distinguished by coercivity alone and a thermal demagnetization experiment should be conducted to search for the ~300°C Curie temperature of greigite.

(a)

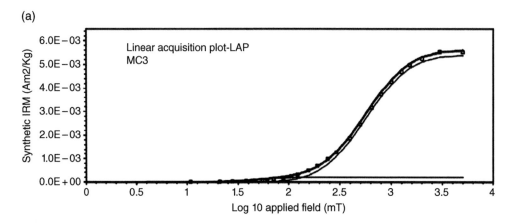

(b)

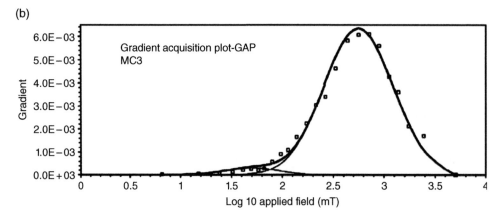

Figure 2.6 IRM acquisition modeling of samples from the Carboniferous Mauch Chunk Formation. (a) The linear acquisition plot shows the raw IRM acquisition data. (b) The gradient acquisition plot shows the fitting of log normal distributions for magnetic particle coercivity distributions to the first derivative of the IRM linear acquisition plot. Source: Bilardello & Kodama 2010.

2.4.2 Magnetic Particle Anisotropy

The reason that paleomagnetism is such an important field of study is that the small ferromagnetic grains in most rocks retain their magnetization in the same direction for geological significant time periods, i.e., billions of years. The magnetic stability of ferromagnetic grains is the result of the magnetization of a grain being fixed in place by the magnetic anisotropy of the grains. Anisotropy means simply that individual magnetic grains are more easily magnetized in one direction, in the grain, than another direction. Magnetic anisotropy can arise from three different causes; the shape of the magnetic grain, the crystallography of the magnetic grain, and the effects of stress deforming the grain. For magnetite, shape is the important type of anisotropy for stabilizing its magnetization. For hematite, crystallography is important in controlling the magnetic grain's stability.

In shape anisotropy, a magnetic particle is more easily magnetized along its long axis, so an ellipsoidal grain will be magnetized along its long axis. Perpendicular to the long axis is the hard magnetization direction, which provides an energy barrier to the magnetization flipping over to the opposite long axis direction. The height of the energy barrier is a function of the microscopic coercivity, h_c, of the grain. In fact, the long-term stability of the magnetic grain is a function of the trade-off between the energy barrier, $vh_c j_s$, holding the magnetization along the easy axis direction and the thermal energy (kT) that can on occasion provide enough energy to flip a grain's magnetization.

$$t = \frac{1}{C} \exp \frac{vh_c j_s}{2kT} \qquad (2.4)$$

where t is the relaxation time, C is a constant called the frequency factor (10^{-8} s^{-1}), v is the volume of the grain, h_c is the grain's microscopic coercivity, j_s is the spontaneous magnetization of the grain, k is the Boltzmann's constant, and T is the temperature. This equation shows that a ferromagnetic grain's size is critical in determining its stability. Magnetite grains as small as 10–20 nanometers have short relaxation times of 100s of seconds and are called superparamagnetic grains. They are ferromagnetic but behave like paramagnetic grains; they line up with an applied field causing an induced magnetization, but their directions randomize when the field is turned off. As ferromagnetic grains increase in size up to 40–50 nanometers, their relaxation times quickly increase to billions of years and they are paleomagnetically stable.

2.4.3   Domain State

As the ferromagnetic grains of magnetite grow larger into the several micron size range, their magnetostatic energy increases. One way of picturing the increase in energy is that as the grain increases in size, more and more of the grain's surface is covered with magnetic "charges" of the same sign. Since charges of the same sign repel each other, the magnetostatic energy of the grain increases. To reduce its overall energy, the grain subdivides into uniformly magnetized subregions, or domains, which are magnetized in opposite directions. There is a critical diameter where the energy required for building the wall or boundary between these magnetic domains is just equal to the energy saved by subdividing into a MD grain. The grain has reduced its overall energy, but the overall magnetization of the grain is also reduced. Domain walls can move to readjust a grain's magnetization to agree, energetically, with an applied field. The domain with a magnetization parallel to the applied field grows at the expense of the domain with a magnetization opposite to the applied magnetic field. MD wall movement and superparamagnetism, mentioned above, are important mechanisms for ferromagnetic susceptibility. The MD grains also are less magnetically stable than the smaller stable SD grains.

Grains that fall into the central region of the Day plot are called PSD grains. Much natural magnetic grain behavior falls into this category. These are small MD grains that behave as SD grains. Various explanations have been offered to account for this behavior. These explanations include (1) that the magnetization of the domain wall separating the domains of a two domain grain gives the grain a net magnetization or (2) that the domain walls are restricted by vacancies in the crystal lattice from positioning themselves to exactly cancel out the grain's overall moment or (3) that the spin moments in adjacent Fe atoms aren't exactly parallel but take on complicated "vortex" or "flower" configurations.

2.5   Environmental Magnetic Parameters

2.5.1   Individual Environmental Magnetic Parameters

In rock magnetic cyclostratigraphy, the main goal is to measure a rock magnetic parameter, throughout a sedimentary sequence, that will have encoded a climate-driven process. Environmental magnetic studies provide many examples of rock magnetic parameters being used as climate proxies, but the best rock magnetic parameters for cyclostratigraphic studies should be fairly easy to measure, because typically 1000s of samples need to be processed, and fairly easy to interpret.

Magnetic susceptibility has been used in many environmental magnetic studies to detect climate variations. It is a straightforward measure of the concentration of magnetic minerals in a sample and is a quick and easy measurement. It has been successful in delineating the loess–paleosol sequences in the Chinese Loess Plateau, which can be directly correlated to the $\delta^{18}O$ climate signal of glacial–interglacial cycles in marine sediments (Liu et al. 2012). A recent rock magnetic cyclostratigraphic study of Plio-Pleistocene marls from the Stirone River section in Italy (Gunderson et al. 2012) shows that susceptibility can be an excellent recorder of Milankovitch cycles in marine sediments. Data compiled by Peters and Dekkers (2003) shows that magnetic susceptibility is not strongly dependent on magnetic particle size for the important natural magnetic minerals. Susceptibility measurements are typically made with a susceptibility meter, using an inductance bridge circuit, that measures the magnetization induced in a sample while an alternating magnetic field with strength similar to the Earth's (50–100 µT) is applied. An alternating field is used to avoid giving the sample a viscous magnetization.

Despite its success in cyclostratigraphic studies, the interpretation of a magnetic susceptibility cyclostratigraphic signal is not straightforward. Magnetic susceptibility measures the concentration variations of not only ferromagnetic minerals, i.e., magnetite, hematite, and Fe sulfides, but also the paramagnetic Fe-rich silicates and diamagnetic quartz and carbonate in a rock. It is preferable to be able to target a specific population of magnetic grains for cyclostratigraphic analysis in order to better understand how climate cycles were encoded. For this reason, measuring the concentration variations of specific ferromagnetic minerals is preferred for cyclostratigraphy studies.

The rock magnetic parameter of choice for many of the earliest rock magnetic cyclostratigraphic studies (Latta et al. 2006; Kodama et al. 2010) is anhysteretic remanent magnetization (ARM). ARM is a laboratory-applied remanence. A sample is placed in an alternating magnetic field that decays to zero over approximately a minute. The initial peak, alternating field used in most laboratories is approximately 100 millitesla (mT). The application of an alternating magnetic field that decays to zero is the procedure used for standard alternating field demagnetization, and modified alternating field demagnetizers are usually used to apply ARMs. The modification involves the ability to apply a small, constant, biasing magnetic field during the ramp-down of the alternating field, so the sample is left with a remanence that is subsequently measured on a rock magnetometer. The constant biasing magnetic fields used for ARM application are usually about the same strength as the Earth's magnetic field (~ 50 μT). Since the strength of ARM is dependent on the strength of the biasing field, many workers normalize the ARM by the biasing field's intensity. The field-normalized ARM is denoted the **ARM susceptibility**, $\chi_{ARM}=ARM/H_b$, where H_b is the biasing field, and allows comparison of the results from different laboratories. The power of using ARM or χ_{ARM} for cyclostratigraphic studies is that it measures the concentration variations of only the ferromagnetic minerals in a sample. Since most laboratories cannot apply ARMs in peak alternating fields higher than about 100 mT, most ARM measurements are limited to lower coercivity ferromagnetic minerals like magnetite, titanomagnetite, or greigite. Paramagnetic and diamagnetic minerals do not contribute to an ARM. Similarly, if very high coercivity magnetic minerals are present in the rock, e.g., hematite or goethite, they will typically not be activated by a 100 mT alternating field and will not contribute significantly to an ARM. While ARM or χ_{ARM} are usually interpreted to measure concentration variations in a sedimentary sequence, χ_{ARM} is strongly dependent on magnetite grain size (Peters & Dekkers 2003). This relationship is to a large extent the result of SD magnetite grains smaller than about 0.1 μm acquiring χ_{ARM}s approximately 10 times stronger than magnetite grains in the 0.1–10 μm size range, where the ARM acquired is not a strong function of grain size. This strong sensitivity of small SD magnetite grains to ARM application should be considered when interpreting ARM rock magnetic cyclostratigraphies. For small SD magnetite grains <0.1 μm in size, χ_{ARM} will be strongly dependent on grain size, but the flat response of χ_{ARM} to grain sizes in the 0.1–10 μm range indicates that for these larger grains χ_{ARM} is more of a concentration-dependent parameter. The magnetic grain size of the magnetic minerals should always be checked with the magnetic parameter ratios discussed below, to see if the ARM variations could be due to grain size variations or primarily due to concentration variations. Either dependence could serve as a climate proxy, so ARM can be a useful rock magnetic cyclostratigraphic parameter in either case.

IRM can measure the concentration variations of high coercivity ferromagnetic grains. IRMs are applied to a rock sample simply by exposing it to a constant (DC) magnetic field. This can be easily done with a large electromagnet, but maximum field strengths only several hundred mT are typically possible. For high field strengths of 1000s of mT, impulse magnetizers are used in which a capacitor is discharged through a coil. Even though the field is only applied for less than a second, it allows efficient application of an IRM in fields as high as 5000 mT that can easily magnetize high coercivity goethite. However, IRM application is a different process than ARM application. In an ARM, the magnetizations of individual magnetic grains are "tickled" into alignment with the biasing magnetic field. An IRM is more of a "sledgehammer" approach with the magnetizations being forced into alignment with the applied field. ARM is thought to mimic the natural thermal remanent magnetization process that magnetizes the magnetic particles in an igneous rock. In a thermal remanence, the high temperatures of the cooling igneous rock reduce the relaxation times of the magnetic grains' magnetization to only minutes or seconds causing a magnetic grain's moment to flip back and forth along the grain's easy magnetization axis. IRM probably mimics the natural magnetization caused by a lightning strike. SD magnetite grains are no more efficient at acquiring an IRM than magnetite grains in the 0.1–10 μm grain size range, so IRMs are not grain size dependent over the 0.01–10 μm grain size range and a good measure of magnetic mineral concentration variations. Magnetite grains larger than 10 μm are highly likely to be MD grains and IRMs show grain size dependence for these larger grains (Peters & Dekkers 2003). IRMs are typically used to magnetize rock samples containing high coercivity hematite for rock magnetic cyclostratigraphic studies. Most ARM application equipment cannot reach the coercivities of hematite. Hematite grains in the 0.1–1 μm grain size range acquire IRMs with slightly less efficiency than larger hematite grains (1–100 μm), so there is a slight grain size dependence that should be considered in the interpretation of IRM rock magnetic cyclostratigraphic data.

The stepwise application of IRMs to a sample in progressively higher fields until its magnetization saturates is an IRM acquisition experiment. The fields at which different magnetic minerals reach saturation is a measure of the magnetic mineral's mean coercivity and can be used for magnetic mineral identification. Magnetite and the ferromagnetic sulfide, greigite, have mean coercivities in the 10s to 100s of mT. Hematite has coercivities in the 100s to 1000s of mT, and goethite has mean coercivities of 1000s of mT. The different magnetization coercivity components used for magnetic mineral identification can be obtained from an IRM acquisition experiment by using Kruiver et al.'s (2001) coercivity analysis modeling routine. The magnetization coercivity components are assumed to be log normal distributions. These log normal distributions are fit to the gradient of the IRM acquisition curve (Figure 2.6).

2.5.2 Ratios of Environmental Magnetic Parameters

2.5.2.1 χ_{ARM}/χ

The χ_{ARM}/χ ratio was first proposed as a magnetic grain size measurement by King et al. (1982). Since single domain magnetite grains typically acquire a strong χ_{ARM} and susceptibility normalizes for magnetic mineral concentration, this ratio depends on the domain state of low coercivity ferrimagnetic grains (magnetite) to measure grain size. High ratios (~5–15) indicate small, SD grains; low ratios (<1–2) are coarser MD grains (Peters & Dekkers 2003). The caveat for using this ratio is that both parameters (χ_{ARM} and χ) must be measuring the magnetization of the same magnetic mineral in the sample. Paramagnetic and diamagnetic mineral contributions to susceptibility will reduce the effectiveness of this ratio as a grain size measure.

Peters and Dekkers's (2003) compilation also shows that SIRM/χ can be used as magnetic grain size measure, but the relationship between grain size and SIRM/χ is not as well defined as the χ_{ARM}/χ ratio, particularly for magnetite, in the critical 0.1–10 µm grain size range.

2.5.2.2 ARM/SIRM (χ_{ARM}/M_{rs})

The ratio of ARM (or ARM susceptibility) to saturation remanence (designated as either SIRM or M_{rs}) is also used to detect magnetic mineral grain size. The advantage this ratio has over the χ_{ARM}/χ ratio is that both ARM and SIRM are remanence measurements, so there is no concern about the contribution of paramagnetic or diamagnetic minerals to the measurements. However, there is the possibility that the ARM and SIRM measurements may activate different magnetic mineralogies in a sample. ARM is preferentially carried by the low coercivity ferrimagnetic minerals (e.g., magnetite and greigite) and SIRM could also be carried by higher coercivity minerals (e.g., hematite and pyrrhotite). The ratio is meaningless as a grain size measure if ARM and SIRM are carried by two different magnetic minerals in the sample, so the magnetic mineralogy should be carefully determined before interpreting ARM/SIRM variations.

2.5.2.3 Goethite to Hematite Ratio

Both goethite (FeOOH) and hematite (Fe_2O_3) can be formed pedogenically and their relative abundance is controlled by moisture availability during soil formation (Yapp 2001; Harris & Mix 2002; Abrajevitch et al. 2009). If both goethite and hematite are detrital minerals and haven't been affected postdepositionally by heating or diagenesis, their ratio could be a powerful proxy for moisture availability (precipitation vs. evaporation) in the source area of a sedimentary rock. Although there are geochemical techniques for measuring the goethite to hematite ratio in a sedimentary rock, magnetic techniques are relatively fast, cheap, and nondestructive. A magnetic ratio can be constructed in order to detect variations in the goethite to hematite ratio in a rock. Goethite is distinguished magnetically by having very high coercivities (1000s of mT) but a Neel temperature of only about 125°C

(Table 2.2). Hematite has coercivities in the 100s to 1000 mT range, but a much higher Neel temperature of 680°C. Therefore, to detect the goethite to hematite ratio, a sample can be given a saturation magnetization (SIRM), probably using field strengths in the 4–5 T range, then alternating field demagnetized at ~100 mT (SIRM$_{100mT}$), to remove the contribution of any ferrimagnetic minerals (e.g., magnetite), and then heated to ~130°C (SIRM $_{100mT\ 130°C}$). Heating to this low temperature should remove the magnetization of any goethite that was activated by SIRM acquisition. The goethite/hematite ratio is determined magnetically by:

$$\frac{G}{H} = \frac{SIRM_{100mT} - SIRM_{100mT130°C}}{SIRM_{100mT}} \qquad (2.5)$$

2.5.2.4 S Ratio and Hard Isothermal Remanent Magnetization (HIRM)

Magnetic parameter ratios can also be used to detect the relative proportions of different magnetic minerals based on their coercivities. The S ratio measures the relative proportion of magnetite to hematite in a sample. It is measured by saturating the remanence of a sample (SIRM), then applying a backfield to the sample in a field strength that should magnetize all the magnetite. Since the theoretical maximum microscopic coercivity for an infinitely long magnetite needle is 300 mT, this value is often used for the backfield. However, smaller field strengths can be used to detect different grain sizes or shapes of the magnetite. Sometimes a range of values is used to see how magnetic behavior changes as the strength of the backfield varies. The S ratio can be calculated in a variety of ways, but the most common formula is:

$$S\,ratio = \frac{-IRM_{-0.3T}}{SIRM} \qquad (2.6)$$

The S ratio varies from +1 where all the magnetic grains in the sample are magnetite (coercivities less than 0.3 T) so that the sample reaches saturation both initially (SIRM) and in the backfield (IRM$_{-0.3T}$). The negative sign ensures that the sample containing mainly magnetite will have a positive ratio. If the sample has only hematite particles, the 0.3 T backfield will not be able to change its initial SIRM and the S ratio will have a value of −1. Intermediate values indicate more or less contributions of magnetite and hematite. The relative magnetizations of magnetite (92 Am^2/kg) and hematite (0.4 Am^2/kg) need to be considered if the S ratio is used to quantitatively estimate the relative proportions of magnetite and hematite. Liu et al. (2012) point out that the variations in the S ratio are nonlinear and nonunique, so it can only give a qualitative estimate of the proportion of low coercivity to high coercivity magnetic minerals.

The results of saturation remanence (SIRM) and the backfield IRM can be combined in a different way to determine the **hard isothermal** remanent magnetization (HIRM) or hard IRM which is used as a measure of the contribution of high coercivity minerals to the IRM. It is calculated by

$$HIRM = \frac{SIRM + IRM_{-0.3T}}{2}$$
(2.7)

2.6 Identification of Magnetic Mineralogies and Choosing a Rock Magnetic Parameter for Cyclostratigraphy

In planning a rock magnetic cyclostratigraphic study, it is important to first identify the magnetic minerals that characterize the sedimentary sequence. Of course, a standard paleomagnetic study would provide that information and also the information about whether the magnetic minerals are primary or not, an absolute necessity for a meaningful cyclostratigraphic study. The coercivity of the magnetic minerals in a rock and reference to Table 2.2 can be used for identification. IRM acquisition measurements of a subset of samples from the rocks will help determine the different coercivity components in a rock. It is not uncommon for sedimentary rock samples to contain more than one magnetic mineral. The rock magnetic parameter for the cyclostratigraphic study can be chosen to target variations in the concentration or grain size of just one of these minerals, particularly if evidence suggests that the mineral is most likely primary.

If magnetite is present, it is likely to be a primary, depositional magnetic mineral, and ARM is a good way of determining concentration/grain size variations throughout the sedimentary sequence. Even if a higher coercivity mineral like hematite or goethite is also present in the rock, it is unlikely that the 100 mT peak alternating field typically used to apply an ARM will activate and magnetize the higher coercivity minerals. If hematite is the main magnetic mineral in a rock, typical of red beds, IRM may be the only way to determine its concentration variations. IRM acquisition experiments can be used to decide the type of DC magnetic field that should be used to apply the IRM. If magnetite and hematite are both present in a rock, IRM that has been alternating field demagnetized at about 100 mT can be used to isolate just the hematite concentration variations.

Susceptibility is a quick and easy measurement and has been used successfully for rock magnetic cyclostratigraphic studies, but it will be important to determine what mineral is contributing, or dominating, the susceptibility measurement since it could be diamagnetic, paramagnetic, and/or ferromagnetic minerals. One aid to identification is to measure the susceptibility of a suite of samples at room temperature and after they have been cooled to liquid nitrogen temperature (77 K) by immersion for a minute or two in liquid nitrogen. Paramagnetic minerals have their magnetization increased by a factor of about 3.8 (from the Curie Law of susceptibility) while ferromagnetic minerals see little change in their susceptibility.

As mentioned earlier and in the case studies (Chapter 6), magnetic parameters can be tailored to the magnetic characteristics of the rock. Both S ratio and the magnetically measured goethite-to-hematite ratio can be useful rock magnetic measurements for cyclostratigraphy.

References

Abrajevitch, A., Van der Voo, R., & Rea, D.K. (2009) Variations in relative abundances of goethite and hematite in Bengal Fan sediments: Climatic vs. diagenetic signals. *Marine Geology, 267,* 191–206. DOI:10.1016/j.margeo.2009.10.010.

Bilardello, D. & Kodama, K.P. (2010) A new inclination shallowing correction of the Mauch Chunk Formation of Pennsylvania, based on high-field AIR results: Implications for the Carboniferous North American APW path and Pangea reconstructions. *Earth and Planetary Science Letters, 299,* 218–227. DOI:10.1016/j.epsl.2010.09.002.

Butler, R.F. (1992) *Paleomagnetism: Magnetic Domains to Geologic Terranes,* 319 pp. Blackwell Scientific Publications, Boston.

Day, R., Fuller, M.D., & Schmidt, P.W. (1977) Hysteresis properties of titanomagnetites: Grain size and composition dependence. *Physics of the Earth and Planetary Interiors, 13,* 260–266. DOI:10.1016/0031-9201(77)90108-X.

Dunlop, D.J. & Ozdemir, O. (1997) *Rock Magnetism: Fundamentals and Frontiers.* Cambridge University Press, Cambridge.

Ellwood, B.B., Tomkin, J.H., Hassani, A.E., Bultynck, P., Brett, C.E., Schindler, E., Feist, R., & Bartholomew, A.J. (2011) A climate-driven model and development of a floating point time scale for the Middle Devonian Givetian Stage: A test using magnetostratigraphy susceptibility as a climate proxy. *Palaeogeography, Palaeoclimatology, Palaeoecology, 304,* 85–95. DOI:10.1016/j.palaeo.2010.10.014.

Evans, M.E. & Heller, F. (2003) *Environmental Magnetism: Principles and Applications of Enviromagnetics,* 299 pp. Academic Press, Amsterdam.

Gillett, S.L. & Van Alstine, D.R. (1982) Remagnetization and tectonic rotation of upper Precambrian and lower Paleozoic strat from the Desert Range, southern Nevada. *Journal of Geophysical Research, 87,* 10929–10953. DOI:10.1029/JB087iB13p10929.

Gradstein, F.M., Ogg, J.G., & Smith, A. (2004) *A Geologic Time Scale.* Cambridge University Press, New York.

Gradstein, F.M., Ogg, J.G., Schmitz, M.D., & Ogg, G.M., eds. (2012) *The Geologic Time Scale 2012,* 1144 pp. Elsevier, Oxford.

Gunderson, K.L., Kodama, K.P., Anastasio, D.J., & Pazzaglia, F.J. (2012) Rock-magnetic cyclostratigraphy for the Late Pliocene-Early Pleistocene Stirone section, Northern Apennnine mountain front, Italy, *Geological Society, London, Special Publications, 373,* 26. DOI:10.1144/SP373.8

Harris, S.E. & Mix, A.C. (2002) Climate and tectonic influences on continental erosion of tropical South America 0-13 Ma, *Geology, 30,* 447–450. DOI:10.1130/0091-7613(2002)030<0447:CATIOC>2.0.CO;2.

Hinnov, L.A., Kodama, K.P., Anastasio, D.J., Elrick, M., & Latta, D.K. (2013) Global Milankovitch cycles recorded in rock magnetism of the shallow marine lower

Cretaceous Cupido Formation, northeastern Mexico. In: Jovane, L., Herrero-Bervera, E., Hinnov, L.A., & Housen, B.A. (eds), *Magnetic Methods and the Timing of Geological Processes*, p. 15. Geological Society, London.

Horng, C.-S. & Roberts, A.P. (2006) Authigenic or detrital origin of pyrrhotite in sediments? Resolving a paleomagnetic conundrum. *Earth and Planetary Science Letters*, *241*, 750–762. DOI:10.1016/j.epsl.2005.11.008.

King, J., Banerjee, S.K., Marvin, J., & Ozdemir, O. (1982) A comparison of different magnetic methods for determining the realtive grain size of magnetite in natural materials: Some results from lake sediments. *Earth and Planetary Science Letters*, *59*, 404–419. DOI:10.1016/0012-821X(82)90142-X.

Kodama, K.P. (2012) *Paleomagnetism of Sedimentary Rocks: Process and Interpretation*, 157 pp. Wiley-Blackwell, Oxford.

Kodama, K.P., Anastasio, D.J., Newton, M.L., Pares, J., & Hinnov, L.A. (2010) High-resolution rock magnetic cyclostratigraphy in an Eocene flysch, Spanish Pyrenees, *Geochemistry, Geophysics, Geosystems*, *11*(6). DOI:10.1029/2010GC003069

Kruiver, P.P., Dekkers, M.J., & Heslop, D. (2001) Quantification of magnetic coercivity components by the analysis of acquisition curves of isothermal remanent magnetisation. *Earth and Planetary Science Letters*, *189*, 269–276. DOI:10.1016/S0012-821X(01)00367-3.

Latta, D.K., Anastasio, D.J., Hinnov, L.A., Elrick, M., & Kodama, K.P. (2006) Magnetic record of Milankovitch rhythms in lithological noncyclic marine carbonates. *Geology*, *34*, 29–32. DOI:10.1130/G21918.1.

Liu, Q., Roberts, A.P., Larrasoana, J.C., Banerjee, S.K., Guyodo, Y., Tauxe, L., & Oldfield, F. (2012) Environmental magnetism: Principles and applications. *Reviews of Geophysics*, *50* (RG4002). DOI:10.1029/2012RG000393

Maher, B.A., Thompson, R., & Hounslow, M.W. (1999) *Quaternary Climates, Environments and Magnetism*. Cambridge University Press, Cambridge.

Mayer, H. & Appel, E. (1999) Milankovitch cyclicity and rock-magnetic signatures of palaeoclimatic change in the Early Cretaceous Biancone Formation of the Southern Alps, Italy. *Cretaceous Research*, *20*, 189–214. DOI:10.1006/cres.1999.0145.

McCabe, C. & Elmore, R.D. (1989) The occurrence and origin of Late Paleozoic remagnetization in the sedimentary rocks of North America. *Reviews of Geophysics*, *27*, 471–494. DOI:10.1029/RG027i004p00471.

Minguez, D., Kodama, K.P., & Hillhouse, J.W. (2014) Paleomagnetic and cyclostratigraphic constraints on the duration of the Shuram carbon isotope excursion, Johnnie Formation, Death Valley region, CA, Geochem. Geophys. Geosys. (in review).

O'Reilly, W. (1984) *Rock and Mineral Magnetism*, 220 pp. Blackie, Glasgow/London.

Peters, C. & Dekkers, M.J. (2003) Selected room temperature magnetic parameters as a function of mineralogy, concentration and grain size. *Physics and Chemistry of the Earth*, *28*, 659–667. DOI:10.1016/S1474-7065(03)00120-7.

Tauxe, L. (2010) *Essentials of Paleomagnetism*, 489 pp. University of California Press, Berkeley.

Tauxe, L., Bertram, H.N., & Seberino, C. (2002) Physical interpretation of hysteresis loops: Micromagnetic modelling of fine particle magnetite. *Geochemistry, Geophysics, Geosystems*, *3*. DOI:10.1029/2001GC000241.

Thompson, R. & Oldfield, F. (1986) *Environmental Magnetism*. Allen and Unwin, London.

Van der Voo, R. & Torsvik, T. (2012) Remagnetization history. In: Elmore, R.D., Muxworthy, A.R., Aldana, M.M., & Mena. M. (eds), *Remagnetization and Chemical Alteration of Sedimentary Rocks*. Geological Society, London.

Woods, S.D., Elmore, R.D., & Engel, M.H. (2002) Paleomagnetic dating of the smectite-to-illite conversion: Testing the hypothesis in Jurassic sedimentary rocks, Skye, Scotland. *Journal of Geophysical Research*, *107*, EPM 2-1–EPM 2-10. DOI:10.1029/2000JB000053.

Yapp, C. (2001) Rusty relics of Earth history: Iron(III) oxides, isotopes, and surficial environments. *Annual Review of Earth and Planetary Sciences*, *29*, 165–199. DOI:10.1146/annurev.earth.29.1.165.

3 Magnetostratigraphy

Abstract: This chapter provides the fundamentals of conducting and interpreting a magnetostratigraphy study of sedimentary rocks. Magnetostratigraphy can provide an average sediment accumulation rate for the sedimentary sequence targeted for a rock magnetic cyclostratigraphy study that will help the investigator identify astronomically forced cycles in the cyclostratigraphy. This chapter covers magnetostratigraphic sampling techniques and strategies, paleomagnetic measurements and demagnetization, plotting of magnetostratigraphic data, and the determination of a reversal stratigraphy from the paleomagnetic data. The chapter ends with a brief discussion of the geomagnetic polarity time scale (GPTS) and techniques for tying the reversal stratigraphy to the GPTS to provide absolute time for the sedimentary sequence being studied.

3.1 Introduction

One of the most important considerations for accurately identifying astronomically forced cycles in rock magnetic cyclostratigraphy is an independent assignment of time to a sedimentary sequence so that the approximate length of any cycles observed can be estimated. For example, Olsen and Kent (1996) used the ages of geologic time scale boundaries (Triassic–Jurassic, Carnian–Norian, Norian–Rhaetian) in the Newark Basin lake sediments to put an approximate age scale on the sequence of lithologically recorded, lake depth van Houten cycles. They found that the McLaughlin cycle, which is observed to bundle the shorter van Houten cycles in the sequence, had a duration of anywhere from 308 to 442 kyr, with a mean duration of 397.7 kyr. They identified this cycle as the 405 kyr long eccentricity cycle. (Actually Olsen and Kent (1996) used 412.885 kyr for long eccentricity from Berger et al. (1992).) Once long

Rock Magnetic Cyclostratigraphy, First Edition. Kenneth P. Kodama and Linda A. Hinnov.
© 2015 John Wiley & Sons, Ltd. Published 2015 by John Wiley & Sons, Ltd.

eccentricity was identified and the sequence of Newark basin lake deposits could be scaled to that time assignment, precession and short eccentricity cycles were recognized as the shorter cycles in the record. Biostratigraphic records can also provide time control for a sedimentary sequence, but magnetostratigraphy allows more precise and accurate age control because geomagnetic polarity reversals are relatively quick geologically ($\sim 10^3$–10^4 years) and are not time transgressive, like many biostratigraphic boundaries. Magnetostratigraphy can provide accurate time calibration for cyclostratigraphic studies, but at a lower resolution than the astronomically forced cycles potentially measured with cyclostratigraphy. Without some kind of independent, absolute time control, a serious pitfall of cyclostratigraphic interpretations is misidentification of astronomically forced cycles, the best example being the Latemar controversy (Chapter 6). Magnetostratigraphy and cyclostratigraphy can and should be used together to achieve precession-scale high-resolution chronostratigraphy.

Magnetostratigraphy is a technique in which paleomagnetic measurements of samples, typically from a sedimentary sequence, are used to determine the polarity intervals that record the polarity of the geomagnetic field at the time of the sediment's deposition. Once the sequence or pattern of polarity intervals is identified, a correlation to the geomagnetic polarity time scale (GPTS) (Gradstein et al. 2004, 2012) assigns absolute time. The polarity interval boundaries provide absolute time tie points throughout the sequence and allow average sediment accumulation rates to be calculated. Magnetostratigraphy has, at best, a 10^4 years time resolution, depending on the reversal rate of the geomagnetic field at the time the rocks were deposited. Over the past ~ 160 Ma, the shortest polarity intervals occurred in the last ~ 20 Ma and about 150 Ma when the geomagnetic field reversed on average up to 5 times/Myr. Further back in time, the reversal rate declines to 0 by the Late Cretaceous and doesn't reverse polarity at all from about 84 Ma to about 126 Ma (Gradstein et al. 2012). The same variability in geomagnetic field reversal rate is also observed in the Proterozoic (Pavlov & Gallet 2010). The GPTS is calibrated mainly by dating of seafloor magnetic anomalies recorded in the world's oceans. Since the oldest extant seafloor is Jurassic in age, the well-determined GPTS only goes back to ~ 170 Ma.

The basic techniques needed to collect and measure the paleomagnetic samples necessary to determine a magnetostratigraphy for a sedimentary sequence are covered in great depth in excellent paleomagnetic textbooks (Butler 1992; Tauxe 2010). Opdyke and Channell's (1996) book gives comprehensive coverage of past magnetostratigraphic studies. We will provide only a basic outline of the steps involved for establishing a magnetostratigraphy and urge the reader to consult the textbooks cited above for more detail.

3.2 Measuring Magnetostratigraphy

3.2.1 Sampling Strategy

The first step is to design a field sampling strategy. If the ultimate goal is to tie the reversal stratigraphy measured for the rocks to the GPTS, it is helpful to know roughly how old the rocks are (i.e., whether Eocene, Cretaceous, or Jurassic), so that the geomagnetic field reversal rate can be approximated. The sediment accumulation rate should be estimated, too, since the objective is to sample a polarity interval at a minimum of three separate horizons. Magnetostratigraphic samples are usually collected from separate stratigraphic horizons, or geologic "instants" of time. The most common technique is to collect about three independently oriented cores, drilled with a gasoline-powered sampling drill, from a separate horizon. The horizons should be spaced as evenly as possible through the stratigraphic section at an interval that would ensure about three horizons in a polarity interval (see Butler 1992). Continental fluvial rocks typically have sediment accumulation rates of ~10 cm/kyr (Sadler 1981). With a reversal rate of 4–5/Myr in the Neogene, this sediment accumulation rate would require sampling intervals between 6 and 8 m. By the same reasoning, the slower sediment accumulation rates of pelagic marine sediments (<1 cm/kyr, Sadler (1981)) would indicate a sampling interval of <0.6 m for the Neogene.

Multiple samples are usually taken at each horizon in order to check for the effects of magnetic overprinting after deposition. Rocks can pick up secondary magnetizations due to chemical changes in the rock growing new magnetic minerals or the acquisition of secondary magnetizations by the primary, depositional magnetic minerals from so-called viscous magnetizations (see Butler (1992) or Tauxe (2010)). The acquisition of secondary magnetizations is usually not homogeneously distributed throughout the rocks, so collecting multiple samples from a horizon often yields at least one sample per horizon that is not completely overprinted or not overprinted at all. If all the samples from a horizon have similar paleomagnetic directions, then it is meaningful to calculate a horizon mean direction, and $N = 3$ is the minimum number of samples needed to conduct the Fisher (1953) statistics typically used to analyze paleomagnetic directional data.

3.2.2 Sample Collection

Paleomagnetic samples are collected typically by drilling 25 mm diameter cores with a gasoline-powered drill and diamond tipped drill bits (Figure 3.1). The cores are oriented using a specially designed orienting tool and a magnetic compass. If drilling is not allowed in the sampling area or it is not practical to haul the drill, gasoline, and cooling water needed

Figure 3.1 Sampling for magnetostratigraphy in the Dolomites with a gasoline powered sampling drill. Only one stratigraphic horizon was sampled at each site for this magnetostratigraphic study. Three cores were collected at each site. Source: Spahn et al. 2013.

for drilling to the sampling locality, oriented hand samples can be collected and cores can be drilled from them back at the laboratory. The advantage of drilling cores in the field is that many independently oriented cores can be collected and easily transported back to the laboratory for trimming and measurement. Oriented blocks have the advantage that no specialized, heavy equipment is needed in the field, but hauling and shipping many blocks (at least three per horizon) back home can be daunting.

3.2.3 Measurement and Demagnetization

Once the oriented samples, either cores or blocks, are back at the laboratory, the cores are trimmed to about 21–22 mm in length so that they can be measured with a rock magnetometer. Most rock magnetometers used today are superconducting and can measure a sample in about a minute or two. To remove any secondary magnetizations that the sample may have picked up since deposition, it must be progressively stepwise demagnetized, in which magnetization is removed from the sample, either by heating or by being exposed to an alternating magnetic field, in small steps of increasingly higher temperatures or fields until the sample is completely demagnetized. The magnetizations that are the hardest to remove, at the highest temperatures or alternating fields, are assumed to be the most ancient magnetizations in the rock. But even the most ancient magnetizations in a rock sample are not necessarily as old as the rock, so other tests must be done to constrain the age of the rock's magnetization. These tests will be described in Section 3.2.4.

Thermal demagnetization is the most widely used technique because experience shows it to be the most effective at removing secondary magnetizations, particularly for very ancient rocks or rocks with magnetizations carried by hematite (red beds). Thermal demagnetization involves heating rock samples in magnetically shielded ovens, cooling them, and measuring the remaining remanence at higher and higher temperature steps until they lose their ferromagnetism. This is the Curie temperature for magnetite (580°C) and the Neel temperature for hematite (680°C). The disadvantage of thermal demagnetization is that the heating can often cause new magnetic minerals to grow during the experiment, obscuring the ancient remanence in the rocks.

Alternating field demagnetization exposes a rock sample to peak alternating fields that then smoothly decrease to zero. The rock is measured after each exposure to the alternating field, the peak alternating field is increased in steps until the rock's magnetization is completely removed. Alternating field demagnetization is used on recent unconsolidated sediments that can't be heated without drying out and physically destroying the sample; however, it is less successful at removing secondary magnetizations from ancient rocks. It is also ineffective for rocks with high-coercivity hematite as the main magnetic mineral, i.e., red beds.

The demagnetization of iron sulfides (greigite) is problematic because heating the samples during thermal demagnetization usually causes the iron sulfide to oxidize to magnetite that carries a secondary magnetization acquired during the demagnetization. Even though the Curie temperature of the iron sulfide is lower than the Curie temperature of any depositional magnetite that may also be in the sediments, the secondary magnetization created by the heating swamps the primary magnetization in the rocks. Alternating field demagnetization can be used on these rocks, but the magnetization isolated should be checked to see if it is primary or secondary.

All progressive demagnetization data, thermal or alternating field, is presented in an orthogonal demagnetization diagram (Zijderveld 1967), also known as a vector component diagram (Butler 1992) or a vector endpoint diagram (Tauxe 2010). These important diagrams may appear complicated at first, but they are a good way of presenting the decay of a three-dimensional vector during demagnetization. They are important to use in any paleomagnetic study because they show, at one glance, the quality of the paleomagnetic data. In typical orthogonal demagnetization diagrams, the solid symbols represent one end of the horizontal component of the paleomagnetic vector at some point during demagnetization. The vector's other end is fixed to the origin. The open symbols are used to depict the endpoints of the vertical component of the paleomagnetic vector (Figure 3.2). When both sets of points trend into the origin of the plot, a stable remanence direction has been isolated from the sample at the highest alternating fields or temperatures needed to totally remove the remanence. Although this remanence direction needs to be checked to see if it is primary or secondary, it does show that the remanence is in all likelihood ancient due to its high stability. The age of the remanence is checked with tests that have become common in paleomagnetic studies and are covered in the following section.

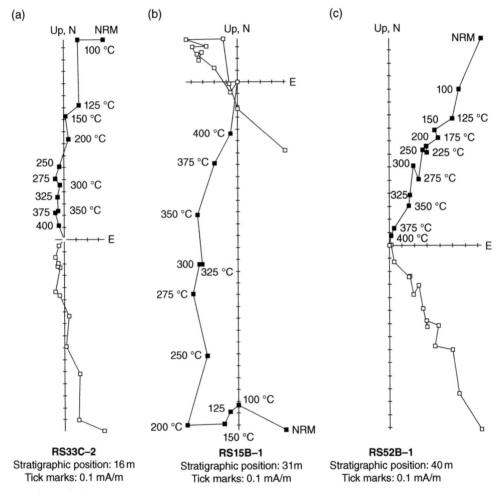

Figure 3.2 Examples of orthogonal demagnetization diagrams (Zijderveld 1967) for thermal demagnetization from a magnetostratigraphic study in Rio Sacuz, Italy. Most of the remanence was removed by thermal demagnetization up to 400°C. Subsequent rock magnetic measurements indicate that the magnetic mineralogy is predominately magnetite. Open symbols are the vertical components and solid symbols are the horizontal components. Data plotted in stratigraphic coordinates. Source: Spahn, Kodama & Preto 2013. Reproduced with permission of John Wiley & Sons, Inc.

3.2.4   Tests to Constrain the Age of Remanence

Once the stable remanence has been isolated in a paleomagnetic sample, it is important to check if the age of the remanence is primary. It is, unfortunately, not uncommon for an ancient magnetization isolated by demagnetization to have been acquired long after the sedimentary rock was deposited. The most important test to constrain the age of remanence is the paleomagnetic fold test (Graham 1949). The magnetizations isolated by demagnetization from samples collected on both limbs of a fold are compared in their *in situ* orientation and also after the beds have been "unfolded" and are mathematically reoriented to be flat-lying. If the magnetizations are better clustered after the beds have been "unfolded," the magnetization passes the fold test and was acquired before the folding. Passage of the fold test does not guarantee that the magnetization is primary, i.e., as old as the rock is, but it does prove that the magnetization is at least ancient and likely to be primary (Figure 3.3). Soft-sediment folds that formed before the sediment was lithified can be used for a fold test and constrain the magnetization to be nearly as old as the rock.

Of course, if the magnetization fails the fold test, it was acquired after folding and is a secondary remagnetization. Statistical tests are used to determine at what level of confidence the scatter of magnetizations can be considered to be pre- or postfolding in age (e.g., McElhinny 1964; McFadden & Jones 1981).

Figure 3.3 Paleomagnetic fold test. The arrows show the mean paleomagnetic magnetization direction in each limb of the fold. (a) The special case when the magnetization is best clustered as the beds are being "unfolded" and indicates a magnetization that was acquired during folding or has been rotated at the grain scale by strain. (b) The geometry of magnetizations that were acquired before folding (prefolding). (c) The failure of the fold test when the magnetization was acquired after folding. Source: Kodama 2012. Reproduced with permission of John Wiley & Sons.

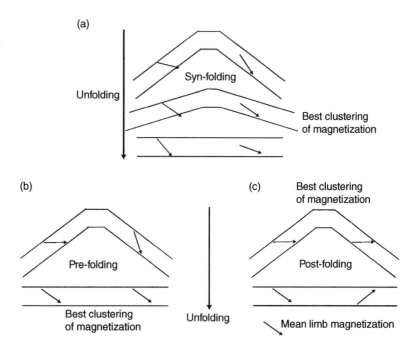

In a conglomerate test (Butler 1992; Tauxe 2010), the magnetizations of individual conglomerate clasts are compared and should be randomly scattered if the magnetization is primary. A blanket remagnetization of the conglomerate unit would produce a magnetization that is uniform. A statistical test of the clast magnetizations checks if the scatter can be considered to be truly random at some level of confidence (e.g., Watson 1956). The conglomerate test is not commonly performed since many sedimentary rocks don't have a conglomerate bed available for the test.

An important test to constrain the age of magnetization for a magnetostratigraphic study is a reversals test. In a simple reversals test, the reversed polarity and normal polarity mean directions are compared to determine if they are antipodal (180° apart) within their 95% confidence limits. When one of the polarity means is rotated 180°, both polarity mean directions should lie within the other polarity mean's 95% confidence limits. The confidence limits are usually calculated with Fisher statistics (Fisher (1953), see Butler (1992)). Non-antipodality of normal and reversed polarity mean directions is interpreted to mean that an incompletely removed overprint, either normal or reversed polarity, has contaminated the results. A more sophisticated application of the reversals test has been developed by McFadden and McElhinny (1990) based on the advances in statistical theory, but basically all versions of the reversals test check whether an overprint persists in the data despite the demagnetization conducted in a standard paleomagnetic study. Pares and Van der Voo (2013) studied the non-antipodality of Tertiary age rocks worldwide and suggested that the shallower inclinations in reversed polarity rocks may be due to the structure of the geomagnetic field rather than a persistent, partially unremoved normal polarity overprint.

In any event, overprinting by secondary magnetizations is a reality for paleomagnetic data and the reversals test is a good way to check for unremoved or incompletely removed overprints.

3.2.5 Plotting of Magnetostratigraphic Data and Determining a Reversal Stratigraphy

To determine a magnetostratigraphy for a sedimentary sequence, the data can be plotted in a variety of ways. In most cases, however, each sample's demagnetized magnetization is converted to a **virtual geomagnetic pole (VGP)**. A VGP is the position of the north geomagnetic pole consistent with the sample's magnetization assuming that the Earth's field is dipolar in geometry. Reversed polarity directions will have their VGPs distributed around the south geographic pole for younger data in which the continents have not strayed far from their present positions. In calculating a VGP position, a magnetization's declination indicates the direction from the site location to the VGP and the magnetization's inclination indicates how far away the VGP is from the site on the surface of the globe. Flat inclinations indicate that the VGP is 90° distant, upward inclinations indicate the VGP is >90° away, assuming the magnetization is normal polarity. Downward inclinations indicate that a normal polarity VGP is <90° away. The regular variation of a dipole field shows that the distance of the VGP (the colatitude) from the site is given by a simple formula:

$$p = \tan^{-1}(2/\tan I) \tag{3.1}$$

where p is the colatitude and I is the sample's paleomagnetic inclination, the angle of the magnetization below the ancient horizontal (bedding for a sedimentary rock).

In the presentation of magnetostratigraphic data, the latitude of the VGP for each sample's magnetization, or the mean of the multiple samples collected from one horizon, is plotted as a function of stratigraphic position. For ancient rocks, in which the continents have moved great distances from their present location, and perhaps twisted around vertical axes, the VGPs for ancient, primary magnetizations will not cluster near the present north and south geographic poles. In that case, it is best to calculate the mean normal polarity VGP for all the data by inverting the reversed polarity directions through the origin. Instead of VGP latitude being plotted as a function of stratigraphic position, the great circle distance of a sample's VGP from the overall VGP mean is plotted as a function of stratigraphic position. This is an estimate of the VGP's ancient paleolatitude (Kent et al. 1995). VGPs that cluster near the north geographic pole (or normal polarity mean VGP) are normal polarity and VGPs that cluster near the south geographic pole are reversed polarity. If a horizon was deposited when the geomagnetic field was in the transition between polarities during a field reversal, the VGP latitude will be at some intermediate position between these two extremes. The real challenge for establishing a magnetostratigraphy is how to determine a horizon's polarity from the latitudes of a horizon's VGPs.

To do this, it is best to establish some rules to be able to determine a reversal stratigraphy. If overprinting by secondary magnetizations is minimal, the establishment of a reversal stratigraphy should be fairly straightforward, but even then the natural behavior of the geomagnetic field will cause some scatter of the data. The paleosecular variation of the geomagnetic field means that VGPs will not be located directly on the north geographic pole, but wander randomly about the north geographic pole. This behavior is shown beautifully in Butler's (1992) textbook. The angular standard deviation of the secular variation over the past 45 Myr has been about 15°, so many workers choose two standard deviations or about 30° from the normal and reversed polarity mean VGPs (or north and south geographic poles, for young data) to assign polarity to a given sample VGP or horizon mean VGP. VGPs that fall outside these bounds could be due to the transition between polarities during a polarity reversal or due to the excursion of the geomagnetic field away from a given stable polarity configuration for a brief (10^3–10^4 years) period. These short excursions appear to be in the continuum of geomagnetic field behavior and examples have occurred in the recent past (e.g., the Mono Lake excursion (Liddicoat & Coe 1979)). During an excursion, a full polarity reversal has not been established (Laj & Channell 2007). Typically, once a horizon's polarity is determined, a stratigraphic column is marked as either normal or reversed polarity, and the pattern of reversed and normal polarity stratigraphic intervals can aid the tie of the sedimentary sequence's reversal stratigraphy to the GPTS (Gradstein et al. 2012).

The addition of significant magnetic overprinting complicates the picture and makes some set of rules necessary. Since the acquisition of secondary magnetizations is typically spatially heterogeneous, it is not uncommon for samples from one horizon to have acquired different degrees of overprinting. Samples from one horizon can have two different polarities. The establishment of "rules" for interpreting the polarity of a horizon in this case lends consistency to the overall polarity stratigraphy interpretation. If the case can be made for dominant normal polarity overprinting, then more weight can be given to reversed polarity samples at a horizon in the determination of the horizon's polarity. This scenario can be the case for younger stratigraphic sections, particularly since the geomagnetic field has been normal polarity for the past 790 kyr and viscous magnetization overprints will be normal polarity. Spahn et al. (2013) were able to make the case that the magnetization of Triassic rocks from the Dolomites were affected by present-day overprinting justifying the heavier weighting of reversed polarity samples in determining a horizon's polarity. Overprinting can cause intermediate directions, in which all the samples at a horizon are partially overprinted and neither reversed or normal polarity, even if the bounds around the normal and reversed polarity VGP positions are expanded to 45° (Kent et al. 2004; Spahn et al. 2013).

Short polarity episodes are observed in the GPTS; therefore, sometimes the finite sampling interval used in any magnetostratigraphic study or the deterioration of the record due to overprinting causes a paleomagnetic study to entirely miss short polarity periods. If only one horizon marks a different polarity from the horizons directly above and below it stratigraphically, it could be due to heterogeneous overprinting affecting just that horizon, or it could be the accurate record of a short polarity interval. Many times the rules established for determining the reversal stratigraphy choose to mark that horizon with less weight, just a "half-bar" in the stratigraphic column showing the polarity interpretation.

An example of the application of rules to establish a magnetostratigraphy can be shown in Spahn et al.'s (2013) study of the stratigraphic section at Rio Sacuz in the Dolomites (Figure 3.4). The rules applied in this example are for a sequence of Triassic age rocks in which present-day overprinting was documented from the demagnetization data. They are as follows:

1 If a horizon contains at least one reversed polarity sample (VGP latitude within 45° of the reversed polarity mean VGP position), the horizon is considered to be reversed polarity.
2 If a horizon contains at least one intermediate direction sample (VGP latitude between 45° from the reversed polarity VGP position and 45° from the normal polarity VGP position), the horizon is intermediate in direction and no polarity can be assigned.

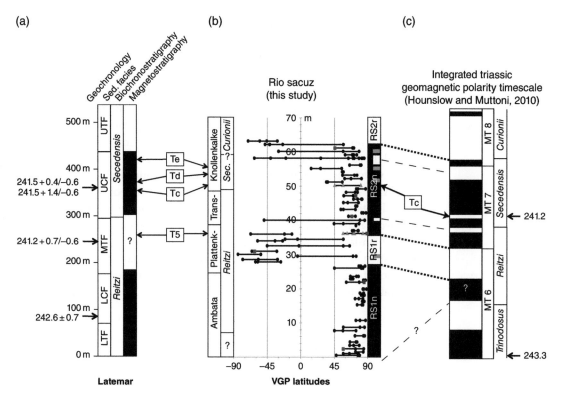

Figure 3.4 Magnetostratigraphy of Triassic carbonate rocks from the Dolomites showing the effects of present-day normal polarity overprinting. (a) Stratigraphic column showing the lithologic interpretations, biostratigraphy, and previous magnetostratigraphy from Kent et al. (2004). (b) VGP latitudes of the samples plotted as a function of stratigraphic position. Note that the VGP latitude for each sample at horizon is plotted and connected with a horizontal line. Red samples are those for which additional rock magnetic data are available. The column immediately to the right shows the local reversal stratigraphy interpretation following the rules indicated in the text. Black is normal polarity, white is reversed polarity. Half bars indicate a polarity interval based on only one horizon. Gray bars are intermediate horizons in which no sample is with 45° either of normal or reversed polarity. (c) The Triassic GPTS and the tie lines between the local magnetostratigraphy and the Triassic GPTS. Source: Spahn, Kodama & Preto 2013. Reproduced with permission of John Wiley & Sons, Inc.

These rules give more weight to any reversed or even intermediate direction samples than to normal polarity samples, because it is assumed that all the samples suffer from varying amounts of normal polarity overprinting. This approach was used by Kodama (1979) for the normally overprinted Plio-Pleistocene Rio Dell Formation in northern California and made sense for the young Rio Dell rocks since they were deposited just before the present Brunhes normal polarity subchron (C1n). This approach only works for the Triassic age rocks from the Dolomites because the present-day geomagnetic field direction is nearly parallel to the Triassic normal polarity direction. In general, there can be no one set of rules for establishing a magnetostratigraphy. A careful paleomagnetic study must be conducted to establish if there is significant secondary magnetic overprinting and how it interacts, geometrically, with the primary magnetization in the rocks. Different situations will require a different set of rules, but rules need to be established to make sure the polarity interpretation is consistently applied through the section.

A technique commonly used for establishing a magnetostratigraphy is to plot the mean direction, or mean VGP latitude, determined for each stratigraphic horizon. Sometimes only the declination or the inclination of the mean paleomagnetic direction is plotted. This approach can work well if there is no significant overprinting, hence reasonably small confidence limits, and the stable demagnetized direction for a sample is easily determined from the vector endpoint diagrams. When using declination or inclination instead of VGP latitude to indicate a sample's polarity, it is important to realize that declination is the best parameter to plot for low-latitude magnetizations that have flat inclinations, and inclination is the best parameter to examine for high-latitude magnetizations, where there can be a large variation in declination, even though the vertical magnetizations may change only a small amount in direction. This caveat is particularly important for magnetostratigraphies of unoriented cores.

3.3 Tying to the GPTS

Once the local reversal stratigraphy has been established, the important final step is to tie it to the GPTS. The GPTS is well established and dated as far back as there is extant ocean floor and is based on the dating of the seafloor magnetic anomalies, e.g., back to 170 Ma (**Chron** M44, (Gradstein et al. 2012)). Tying the reversal stratigraphy to the GPTS allows absolute ages to be assigned to polarity interval boundaries and hence the establishment of the sediment accumulation rate for the sedimentary sequence under study. This in turn allows the determination of the duration of any cycles found in the cyclostratigraphy.

Tying to the GPTS can be tricky, because sediment accumulation rates can change throughout a section, unrecognized hiatuses can remove part of the record, and very short events may be missed because of either small hiatuses in sedimentation or the sampling interval that was used. The stratigraphic sampling interval also affects the resolution of the stratigraphic position of the polarity interval boundaries and hence the calculation of the sediment accumulation rate. One key piece of information that must be established is an approximate age for the stratigraphic section to aid the tie to the GPTS. Biostratigraphy is often used for this purpose. Of course, geochronology of ash layers can also be an important way to determine the age of the sedimentary section. In the Spahn et al. (2013) magnetostratigraphic study in the Dolomites (Figure 3.4), both biostratigraphy and geochronology of ash layers provided a sound tie to the integrated GPTS for the Triassic (Hounslow & Muttoni 2010).

Before about 170 Ma, the GPTS is not continuous and not as well developed or calibrated. The integrated GPTS for the Triassic (252.5–201.6 Ma) developed by Hounslow and Muttoni (2010) is based on the biostratigraphic correlation of various continental sections globally. Hounslow and Muttoni's (2010) Triassic GPTS is at some odds with Kent et al.'s (1995) astronomically tuned GPTS from Newark Basin drill cores for the Late Triassic, so Gradstein et al. (2012) present two options for the Late Triassic GPTS, one based on the astronomically tuned record (Kent et al. 1995) and one based on the biostratigraphically correlated marine sections (Hounslow & Muttoni 2010). The Triassic GPTS has a time resolution, at best, of 20–30 kyr for its magnetozones (polarity intervals). Kent and Olsen (2008) have been able to extend the Newark Basin magnetostratigraphy from 202 to 199.4 Ma using astronomically tuned reversals recorded in Hartford Basin rocks. The period between the end of the seafloor anomaly-based GPTS at around 170 Ma and the Hartford Basin magnetostratigraphy at 199.4 Ma is based on assorted biostratigraphically correlated continental sections from England, France, Switzerland, Spain, and Austria (Gradstein et al. 2004). Gradstein et al. (2012) have compiled a GPTS for the Permian. The field remained in the reversed polarity state for much of the Permian from about 300 Ma until about 267 Ma with perhaps two short normal events at about 286 and 297 Ma. Recent work by Opdyke et al. (2014) suggests that the base of this long polarity interval may be as old as 318 Ma. This long polarity interval is the Kiaman reversed polarity superchron, and like the Cretaceous normal polarity, superchron (84–126 Ma) is a period in Earth history when magnetostratigraphy is impossible. Recently, Pavlov and Gallet (2005) provide evidence that a third superchron occurred in the Ordovician, between about 480 and 460 Ma. The Carboniferous GPTS is fairly continuous and is based on biostratigraphically correlated continental sections from 299 to 359 Ma (Gradstein et al. 2004, 2012), most notably the Late Mississippian (Serpukhovian, 323–331 Ma) magnetostratigraphy compiled from Appalachian Basins (e.g., Mauch Chunk Fm., (DiVenere & Opdyke 1991)) or the Canadian Maritimes (Opdyke et al. 2000). G. Ogg's chart (Ogg 2012) that summarizes the Geologic Time Scale 2012 (Gradstein et al. 2012) provides a quick view of the status of the GPTS for Phanerozoic rocks (Figure 3.5).

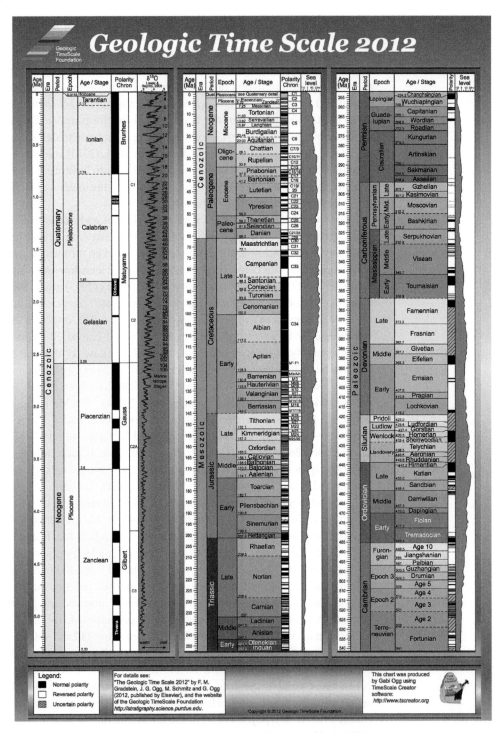

Figure 3.5 Summary of the Geological Time Scale including the GPTS (black and white stripes). The GPTS is reasonably well developed and continuous back to the Carboniferous–Devonian boundary (~350 Ma). Source: Gradstein et al. 2012. Reproduced with permission of Elsevier and Gabi Ogg.

For the Devonian and earlier, the GPTS is not well determined or continuous, and a tie to the GPTS for calibrating cyclostratigraphy is not possible. Only estimates of the duration of a sedimentary sequence are possible assuming the range in reversal rates observed over the well-constrained 0–170 Ma period (from ~5 to 0 reversals/Myr; Merrill et al. 1996) and the number of reversals observed in the section. For example, if for a sequence of sedimentary rocks older than 350 Ma five polarity intervals are observed, the minimum duration for the deposition of the sequence would be estimated to be 1 million years. This is the approach that had to be taken for a Neoproterozoic magnetostratigraphy and rock magnetic cyclostratigraphy conducted by Minguez et al. (2014) and discussed in Chapter 6.

3.4 Providing the Best Time Resolution from Magnetostratigraphy

The main objective of establishing a magnetostratigraphy for the sedimentary sequence targeted for rock magnetic cyclostratigraphy is to have absolute time tie points in the section to get the best estimate of the sediment accumulation rate. An accurate sediment accumulation rate allows the most accurate determination of the average duration for any cycles observed by the cyclostratigraphic measurements. The best estimate of the sediment accumulation rate depends on pinning down as accurately as possible the stratigraphic position of polarity interval boundaries. One way to do this is to use an iterative process. A reconnaissance magnetostratigraphy allows a first-order tie to the GPTS. Once the approximate positions of the polarity boundaries are known, the stratigraphic section can be resampled near to those positions at very close stratigraphic intervals in order to pin down the polarity boundary locations with the highest possible stratigraphic resolution. This approach was used for the development of a rock magnetic cyclostratigraphy for the Eocene Arguis Formation from the Spanish Pyrenees (Kodama et al. 2010) and is detailed in Chapter 6.

References

Berger, A., Loutre, M.F., & Laskar, J. (1992) Stability of the astronomical frequencies over Earth's history for paleoclimate studies. *Science, 255*, 560–566. DOI:10.1126/science.255.5044.560.

Butler, R.F. (1992) *Paleomagnetism: Magnetic Domains to Geologic Terranes*, 319 pp. Blackwell Scientific Publications, Boston.

DiVenere, V.J. & Opdyke, N.D. (1991) Magnetic polarity stratigraphy in the uppermost Mississippian Mauch Chunk Formation, Pottsville, Pennsylvania. *Geology, 19*, 127–130. DOI:10.1130/0091-7613(1991)019<0127:MPSITU>2.3.CO;2.

Fisher, R.A. (1953) Dispersion on a sphere. *Proceedings of the Royal Society of London: Series A, 217*, 295–305. DOI:10.1098/rspa.1953.0064.

Gradstein, F.M., Ogg, J.G., & Smith, A. (2004) *A Geologic Time Scale*. Cambridge University Press, New York.

Gradstein, F.M., Ogg, J.G., Schmitz, M.D., & Ogg, G.M., eds. (2012) *The Geologic Time Scale 2012*, 1144 pp. Elsevier, Oxford.

Graham, J.W. (1949) The stability and significance of magnetism in sedimentary rocks. *Journal of Geophysical Research*, *54*, 131–167. DOI:10.1029/JZ054i002p00131.

Hounslow, M.W. & Muttoni, G. (2010) The geomagnetic polarity timescale for the Triassic: Linkage to stage boundary definitions. *Geological Society, London, Special Publications*, *334*, 61–102. DOI:10.1144/SP334.4.

Kent, D.V. & Olsen, P.E. (2008) Early Jurassic magnetostratigraphy and paleolatitudes from the Hartford continental rift basin (eastern North America): Testing for polarity bias and abrupt polar wander in association with the central Atlantic magmatic province. *Journal of Geophysical Research*, *113*. DOI:10.1029/2007JB005407.

Kent, D.V., Olsen, P.E., & Witte, W.K. (1995) Late Triassic-earliest Jurassic geomagnetic polarity sequence and paleolatitudes from drill cores in the Newark rift basin, eastern North America. *Journal of Geophysical Research*, *100*, 14965–14998. DOI:10.1029/95JB01054.

Kent, D.V., Muttoni, G., & Brack, P. (2004) Magnetostratigraphic confirmation of a much faster tempo for sea-level change for the Middle Triassic Latemar platform carbonates. *Earth and Planetary Science Letters*, *228*, 369–377. DOI:10.1016/j.epsl.2004.10.017.

Kodama, K.P. (1979) New paleomagnetic results from the Rio Dell Formation, California. *Geophysical Research Letters*, *6*, 253–256. DOI:10.1029/GL006i004p00253.

Kodama, K.P. (2012) *Paleomagnetism of Sedimentary Rocks: Process and Interpretation*, 157 pp. Wiley-Blackwell, Oxford.

Kodama, K.P., Anastasio, D.J., Newton, M.L., Pares, J., & Hinnov, L.A. (2010) High-resolution rock magnetic cyclostratigraphy in an Eocene flysch, Spanish Pyrenees. *Geochemistry, Geophysics, Geosystems*, *11*. DOI:10.1029/2010GC003069.

Laj, C. & Channell, J.E.T., eds. (2007) *Geomagnetic Excursions*, pp. 373–407. Elsevier, Amsterdam.

Liddicoat, J.C. & Coe, R.S. (1979) Mono Lake geomagnetic excursion. *Journal of Geophysical Research*, *84*, 261–271. DOI:10.1029/JB084iB01p00261.

McElhinny, M.W. (1964) Statistical significance of the fold test in paleomagnetism. *Geophysical Journal of the Royal Astronomical Society*, *8*, 338–340. DOI:10.1111/j.1365-246X.1964.tb06300.x.

McFadden, P.L. & Jones, D.L. (1981) The fold test in paleomagnetism. *Geophysical Journal of the Royal Astronomical Society*, *67*, 53–58. DOI:10.1111/j.1365-246X.1981.tb02731.x.

McFadden, P.L. & McElhinny, M.W. (1990) Classification of the reversals test in paleomagnetism. *Geophysical Journal International*, *103*, 725–729. DOI:10.1111/j.1365-246X.1990.tb05683.x.

Merrill, R.T., McElhinny, M.W., & McFadden, P.L. (1996) *The Magnetic Field of the Earth, Paleomagnetism, the Core, and the Deep Mantle*. Academic Press, San Diego.

Minguez, D.A., Kodama, K.P., & Hillhouse, J.W. (2014) Paleomagnetic and cyclostratigraphic constraints on the duration of the Shuram carbon isotope excursion, Johnnie Formation, Death Valley region, CA, Geochem., Geophys., Geosys., (in review).

Ogg, G.M. (2012) Geologic time scale 2012. In: *Time Scale Creator*.

Olsen, P.E. & Kent, D.V. (1996) Milankovitch climate forcing in the tropics of Pangea during the Late Triassic. *Palaeogeography, Palaeoclimatology, Palaeoecology, 122*, 1–26. DOI:10.1016/0031-0182(95)00171-9.

Opdyke, N.D. & Channell, J.E.T. (1996) *Magnetic Stratigraphy*, 346 pp. Academic Press, San Diego.

Opdyke, N.D., Roberts, J., Claoue-Long, J., Irving, E., & Jones, P.J. (2000) Base of the Kiaman: Its definition and global stratigraphic significance. *Geological Society of America Bulletin, 112*, 1315–1341. DOI:10.1130/0016-7606(2000)112<1315:BOT KID>2.0.CO;2.

Opdyke, N.D., Giles, P.S., & Utting, J. (2014) Magnetic stratigraphy and palynostratigraphy of the Mississippian-Pennsylvanian boundary interval in eastern North America and the age of the beginning of the Kiaman. *Geological Society of America Bulletin*, 126, 1068–1083, DOI: 10.1130/B30953.

Pares, J. & Van der Voo, R. (2013) Non-antipodal directions in magnetostratigraphy; an overprint bias?. *Geophysical Journal International, 192*, 75–81. DOI:10.1093/gji/ggs027.

Pavlov, V. & Gallet, Y. (2005) A third superchron during the Early Paleozoic. *Episodes, 28 (2)*, 78–84.

Pavlov, V. & Gallet, Y. (2010) Variations in geomagnetic reversal frequency during the Earth's middle age. *Geochemistry, Geophysics, Geosystems, 11*. DOI:10.1029/2009GC002583.

Sadler, P.M. (1981) Sedimentation rates and the completeness of stratigraphic sections. *Journal of Geology, 89*, 569–584. DOI:10.1086/628623.

Spahn, Z.P., Kodama, K.P., & Preto, N. (2013) High-resolution estimate for the depositional duration of hte Triassic Latemar Platform: A new magnetostratigraphy from basinal sediments at Rio Sacuz, Italy. *Geochemistry, Geophysics, Geosystems, 14*, 1245–1257. DOI:10.1002/ggge.20094.

Tauxe, L. (2010) *Essentials of Paleomagnetism*, 489 pp. University of California Press, Berkeley.

Watson, G.S. (1956) A test of randomness of directions. *Monthly Notices of the Royal Astronomical Society, Geophysical Supplements, 7*, 160–161. DOI:10.1111/j.1365-246X.1956.tb05561.x.

Zijderveld, J.D.A. (1967) AC demagnetization of rocks: Analysis of results. In: Collinson, D.W., Creer, K.M., & Runcorn, S.K. (eds), *Methods in Paleomagnetism*, pp. 254–286. Elsevier, Amsterdam.

4

Time Series Analysis for Cyclostratigraphy

Abstract: This chapter reviews the time series analysis of cyclostratigraphy and stresses the detection of variations potentially associated with astronomical forcing. Methods of interpolation, smoothing, filtering, and spectrum estimation are presented, with special emphasis on the Thomson multitaper estimator. Synthetic and real-data examples that use basic MATLAB® functions and custom scripts accompany the presentations. This is followed by an introduction to noise modeling and hypothesis testing. Time-frequency analysis of nonstationary data is addressed with evolutionary spectrograms and complex signal analysis. Spectral coherency estimation with worked examples relevant to astronomical forcing concludes the chapter.

4.1 Introduction

Digital signal processing techniques that were developed for electrical and communications engineering and geophysics (seismology) make up the toolkit that is used today to analyze geological time series. Typical procedures involve preprocessing, spectrum estimation, time-frequency analysis, and correlation. Preprocessing prepares a time series for more advanced procedures such as spectrum estimation, and can involve interpolation, resampling, detrending, and filtering. Deep-time data requires tuning to convert the proxy timescale of the independent variable, usually stratigraphic thickness, into a true timescale. Time-frequency methods such as spectrograms, wavelet analysis, and complex demodulation are used to study the spectral properties of a time series. Coherency and cross-phase spectral analysis measures the correlation of two time series as a function of frequency.

This chapter provides a step-by-step guide to the main procedures that are currently used for the analysis and modeling of cyclostratigraphic time series. The presentation is tailored to the special problems of truncated data

Rock Magnetic Cyclostratigraphy, First Edition. Kenneth P. Kodama and Linda A. Hinnov.
© 2015 John Wiley & Sons, Ltd. Published 2015 by John Wiley & Sons, Ltd.

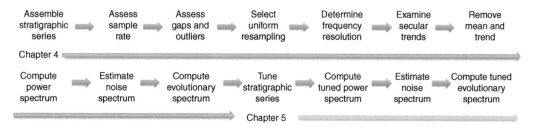

Figure 4.1 Flow chart of the measurement, analysis, and modeling of time series in cyclostratigraphy discussed in Chapters 4 and 5.

series (i.e., finite length) sampled from the stratigraphic record as opposed to the engineering perspective that emphasizes controlled signals. The typical processing and analysis flow (Figure 4.1) show that for investigations of geological phenomena, key decision-making begins at the data collection stage, with selection of the dependent and independent variables.

The time series analysis procedures presented in this chapter are demonstrated on synthetic signals and rock magnetic data collected from the Late Eocene Arguis Formation, Pyrenees, Spain (Kodama et al. 2010), using MATLAB functions and custom scripts (Appendix). The philosophy behind each procedure is discussed together with directions for how to apply the procedures in MATLAB.

4.2 Geological Time Series

Digital time series of geological processes occur in two broad classes: "instrumental-time" or "near-time" and "deep-time." Near-time data are measures of processes that are ongoing or that occurred in the recent past, with the independent variable referred to a controlled timekeeper, e.g., mean-hourly air temperature from a thermometer. Deep-time data are samples of processes that have been recorded in geological materials, e.g., magnetic mineral deposition in deep-sea sediment, in which the independent variable is referred to a proxy timescale that is spatial (core depth or stratigraphic thickness).

Geological time series are further categorized according to the type of process that they represent. Continuous processes evolve through time steadily, supported by a background (continuum) level of random (stochastic) variation, and often accompanied by a comparatively prominent foreground of nonrandom (deterministic) cycling. Discontinuous processes are characterized by intermittent variation, i.e., sudden, short-lived "events" separated by long intervals of time of no action. Here we focus on the measurement and analysis of continuous processes, although significantly, techniques for the analysis of discontinuous processes are now in resurgence (e.g., Hinnov et al. 2012; Olson et al. 2013) (Figure 4.2).

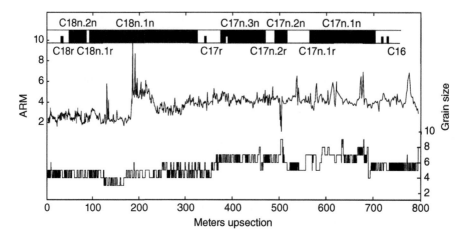

Figure 4.2 Examples of continuous and discontinuous stratigraphic series from the Eocene Arguis Formation, Pyrenees, northern Spain. Discontinuous magnetochrons identified in this section are shown at the top (Kodama et al., 2010); middle: continuous anhysteretic remanent magnetism (ARM) series; bottom: discontinuous grain size series. The ARM series will be used throughout Chapters 4 and 5 to demonstrate the time series tools that are typically used in cyclostratigraphy.

4.3 Time Series Analysis Tools and Eocene Arguis Rock Magnetic Cyclostratigraphy

4.3.1 Sampling and Interpolation

Sample rate is a basic issue with natural data, especially deep-time Earth data with uncertain timescales (for which stratigraphic thickness is the independent variable), or data that cannot be or have not been collected at strict uniform spacings. This is the case with the Arguis ARM series, which was sampled in the field at close spacings of $\Delta d = 20$ cm at the base of the formation, where lithologic cycles were relatively thin, then adjusted to $\Delta d = 75$ cm as bedding became thicker up section, and finally $\Delta d = 1.5$ m for the final 200 m (Figure 4.3a). At various points along the formation, there were gaps (covered outcrop); at other levels, local changes in sampling had to be made. Thus, even at this early stage in the analysis, the practitioner is often confronted by the problem of a nonuniformly sampled series.

There are several options available at this point, for example, windows-based applications SPECTRUM (Schulz & Schaltegger 1997) and for larger datasets, REDFIT (Schulz & Mudelsee 2002) use Lomb–Scargle spectral analysis for nonuniformly sampled time series. Alternatively, the nonuniformly sampled time series can be resampled to a uniform spacing, which allows application of many powerful time series methods that were developed for uniformly sampled data (Thomson 2009). This is the approach taken in our analysis of the Arguis ARM data. However, care needs to be taken to understand the limiting effects of the original

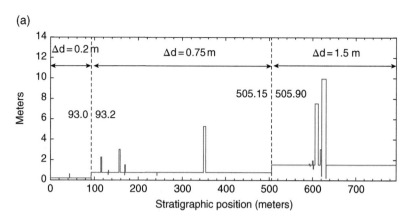

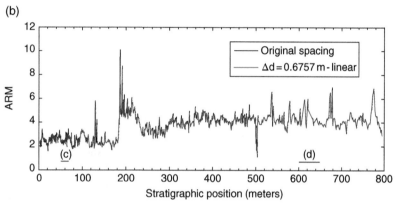

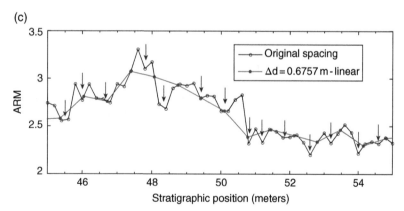

Figure 4.3 Nonuniform sampling and interpolation schemes of the Arguis ARM series in Figure 4.2. (a) Sample rate for ARM along the Arguis formation, showing three changes from $\Delta d = 0.2$ m from 0 to 93 m, $\Delta d = 0.75$ m from 93.2 to 505.15 m, and $\Delta d = 1.5$ m from 505.9 to 800 m. The upward spikes indicate gaps treated here as nonuniform sampling and downward spikes indicate local nonuniform sampling decisions based on outcrop conditions. (b) Original sampled ARM series (black curve) compared with a uniform $\Delta d = 0.6757$ m sampled series (red curve), computed with linear interpolation using *interp1.m* (see Appendix). The horizontal line labeled (c) refers to the detail shown in (c) between 45 and 55 m, showing that the interpolation has missed high-frequency variability (indicated by blue arrows). (d) Detail between 600 and 650 m for an interpolation computed with spline interpolation and $\Delta d = 0.05$ m using *interp1.m*, showing a large artifact introduced between 620 and 630 m (indicated by blue arrow). Such artifacts can be reduced by defining a larger Δd and can be avoided altogether by using linear interpolation.

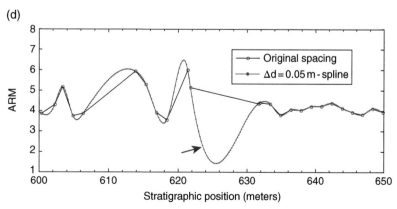

nonuniform sampling on spectral estimates (Section 4.3.5). If hypothesis testing is to be undertaken with null models based on autoregressive noise, then the average spacing must be used, or procedures for nonuniformly sampled time series.

Common pitfalls of interpolation are illustrated in Figure 4.3. The Arguis ARM series has an average spacing of 0.6757 m across the entire 800 m of section. However, if this interval is adopted as the uniform sample rate in an interpolation, some segments of the series will be severely undersampled (Figure 4.3c). Using spline fits are not recommended, as these can produce large perturbations in the interpolation, especially where there are sudden changes in direction over wide gaps (Figure 4.3d). For the demonstrations below, the Arguis ARM stratigraphic series has been linearly interpolated to a uniform spacing of $\Delta d = 0.05$ m, which captures practically all variability with minimal introduction of artifacts.

4.3.2   Detrending, Smoothing, and "Prewhitening"

Frequently, variations and cycles of interest are riding atop a long-term secular trend, which can be extremely high amplitude and quasiperiodic. Estimation and removal of such trends can be crucial for spectral analysis, i.e., to avoid spectral leakage from these high-power components into frequencies of interest. Simple moving averages and weighted mean ("lowess," "loess," or "robust loess") averages may be estimated and removed, in a process called "prewhitening," which will reduce interference of these trends with the rest of the estimated spectrum. This requires experimentation by the practitioner to identify the most appropriate averaging (smoothing) window. Comparison of three 80-m smoothing approaches in Figure 4.4 shows that the robust loess smoothing has picked up on the ~75 m cycling, which may be related to orbital eccentricity cycling (Chapter 5). Thus, if the robust loess smoothing is used for the removal of the long-term trend, signal power could be affected. The large "step-up" in the data series at 182–185 m introduces an irregularity that could interfere with spectral estimates of the low frequencies. Removing or reducing it would diminish its obstructive effects. We selected the lowess smoothing depicted in Figure 4.4 to remove the irregular long-term trend from the Arguis series.

4.3.3   Filtering Basics

Filters are essential tools for isolating specific frequency components in geologic data series for detailed examination. The ideal filter is the "brick wall" filter, with four basic designs: lowpass, highpass, bandpass, and notch. Figure 4.5 shows the basic parameters that are considered in digital filter design, which are constrained to filter resolvable frequencies

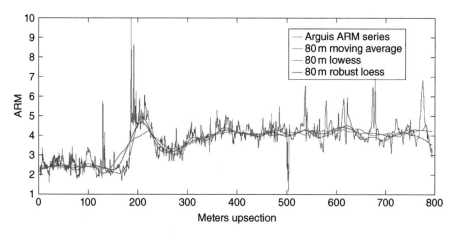

Figure 4.4 Smoothing with simple moving average (green curve), lowess (blue curve), and robust loess (grey curve) methods applied to the Arguis ARM stratigraphic series (red curve) that has been linearly interpolated to $\Delta d = 0.05$ m (as in Figure 4.3), over an 80 m window and using MATLAB's *smooth.m* (smooth(data,80/0.05,'moving'), smooth(data,80/0.05,'lowess'), and smooth(data,80/0.05,'rloess')). The depicted lowess curve will be subtracted from the Arguis series prior to spectral analysis (shown in Figure 4.21).

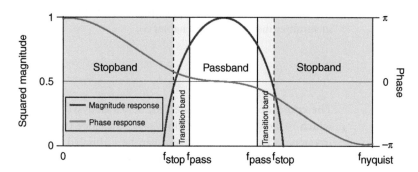

Figure 4.5 Ideal filter compared with practical filter design for a digital signal. The diagram is in the frequency domain from $f = 0$ to $f_{nyquist}$ and shows an ideal passband between two "f_{pass}" frequencies and two "f_{stop}" frequencies designating frequencies passing >50% power. Shaded areas in the stopbands do not pass significant power. The blue curve shows the magnitude response of a typical digital filter with this passband; the green curve shows the filter's phase response, which for cyclostratigraphy needs to be zero in the passband. As a rule, digital filters (such as those provided in MATLAB in the time domain) do not have a zero-phase response (unless applied forward and backward, e.g., using *filtfilt.m*).

within the Nyquist frequency range, i.e., $[0, f_{nyquist} = 1/(2\Delta t)]$. Low-pass filtering ("smoothing") is a common application in cyclostratigraphy and can be conducted with simple or weighted averages (Section 4.3.2) as well as with the digital filters (Section 4.3.4). Band-pass filtering is also commonly applied to extract signals embedded in cyclostratigraphic sequences that are related to eccentricity, obliquity, and precession forcing. Filtering played a major role in the astronomical tuning of the Arguis Formation (Chapter 5).

4.3.4   Digital Filters

Frequency domain filters have proven to be the most effective in cyclostratigraphy. The data are Fourier transformed into the frequency domain, then multiplied by the filter, and inverse-Fourier transformed back to the time domain. The filters must be continuous and differentiable to avoid generating artifacts in the data when transforming back to the time domain. In addition, the filters cannot impose a time delay, i.e., they must be "zero-phase" filters. Here we discuss the Gaussian (Klapper & Harris 1959) and Taner (Taner 2000) filters that perform extremely well in cyclostratigraphy.

The frequency and phase responses of these two filters (Figure 4.6) indicate that strict zero phase is maintained through the passband. The Gauss filter passes power outside of the edges of the cutoff frequencies, whereas the Taner filter has a steep slope. Application of the two filters shows excellent recovery of precession index variation from the 4-myr-long insolation series (Figure 4.7); differences are most noticeable where the amplitude modulations of the precession are low, i.e., when the local frequency of the precession is furthest from the average precession frequency. Both filters faithfully reproduce the phasing of the precession index, although it should be noted that the filtered series are in opposite polarity to the precession as is expected in summer insolation (see Chapter 5).

4.3.5   Spectral Analysis

The cornerstone of univariate time series analysis is the power spectrum, which is the distribution of time series variance (power) as a function of frequency. The power spectrum can be estimated with a variety of nonparametric (Fourier based) or parametric (model based) techniques; each "spectrum estimator" has its own statistics, accuracy (bias), and resolution issues. Spectrum estimators assume stationarity, i.e., the basic statistics of the time series, for example, the mean and the variance, do not change over time. Stationarity is generally not valid for natural time series, especially cyclostratigraphy, which can undergo significant changes in accumulation rate and proxy behavior within a single section. The usual remedy is to apply a spectral estimator as an evolutionary or "running" application through the time series (Section 4.3.7).

Three nonparametric (nonmodel based) spectrum techniques are discussed: the smoothed periodogram, Blackman–Tukey (BT) correlogram, and Thomson multitaper estimator. In particular, the multitaper technique—which includes a harmonic line (amplitude) spectral estimator, an average power spectral estimator, coherency and cross-phase estimators, and other higher-order spectral estimators—provides a "full-service" toolbox of flexible, high resolution, statistically robust estimators that outperform the other nonparametric techniques.

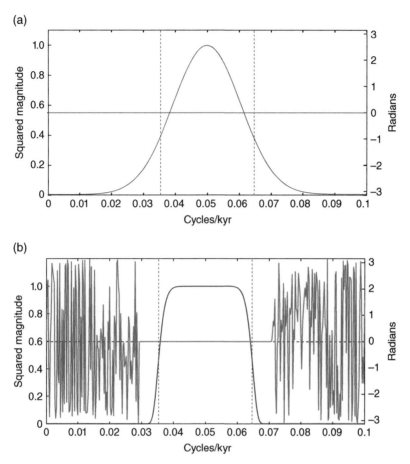

Figure 4.6 Magnitude (blue curves) and phase (green curves) responses ("Bode plots") of two frequency domain **band-pass filters** with equivalent passbands defined with cut-off frequencies at $f_{low} = 0.035$ cycles/kyr and $f_{high} = 0.065$ cycles/kyr (red dashed lines) to recover the precession index from a 65° North summer insolation series sampled at $\Delta t = 1$ kyr, i.e., defining a frequency range of 0 to 0.5 cycles/kyr. (a) Gauss band-pass filter and (b) Taner band-pass filter. These responses were estimated by passing an impulse function (a series with a 1 followed by a string of 0s for a total length of 4096 points) through *gaussfilter.m* and *tanerfilter.m* (Appendix) and then by Fourier transforming the output and computing the squared modulus and phase.

4.3.5.1   The Fourier Transform

The Fourier transform maps the variability of a time series into a set of sines and cosines of different frequencies, each weighted with a "Fourier coefficient." It estimates the frequencies and amplitudes of sinusoids that can be used to reconstruct the time series, thus deconstructing the time series into its "component" periodicities. The phase is determined by the relative contributions of sine and cosine at a given frequency. The **fast Fourier transform (FFT)**, developed for digital computers (Cooley & Tukey 1965) restricts the number of frequencies that are calculated to N/2 for a discrete time series of length N and sampled at Δt intervals. The lowest measurable frequency (longest period) is $1/N\Delta t$, i.e., which has a period that is the

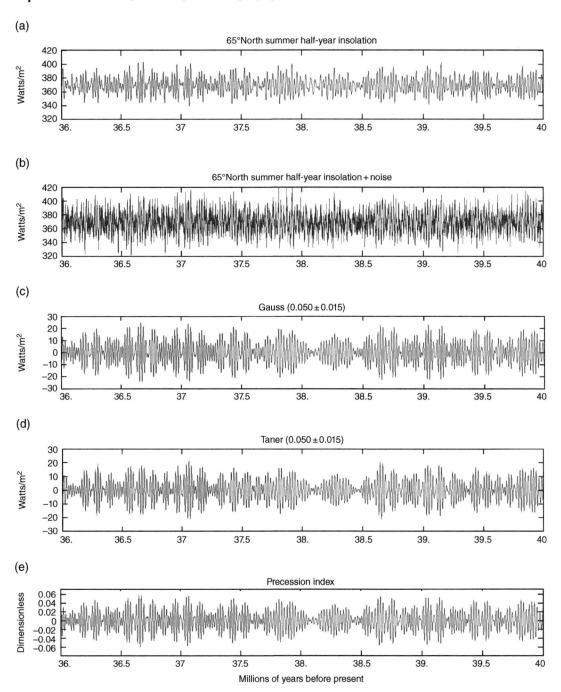

Figure 4.7 Comparison of Gauss and Taner filter performance in extracting a precession index signal from 65° North summer half-year insolation (La2004 model; Chapter 5) sampled at Δt = 1 kyr from 36 to 40 Ma. (a) Summer half-year mean insolation for 65°North, calculated with *Analyseries 2.0.4*. (b) Same as (a) plus white Gaussian noise with variance equal to that of the insolation. (c) Gauss band-pass filtered series of the precession band insolation. (d) Taner band-pass filtered series of the precession band insolation. (e) The actual precession index used to compute (a). The applied filters were defined with the parameters indicated in Figure 4.6. The precession-filtered summer insolation has opposite polarity to the precession index (see Chapter 5).

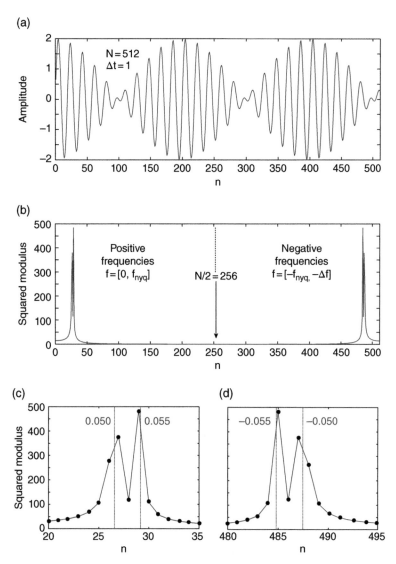

Figure 4.8 (a) Real-valued digital time series of length N = 512 with two frequencies, 0.05 and 0.055, $y_n = \sin(2\pi n\Delta t 0.050) + \sin(2\pi n\Delta t 0.055)$, n = 1,2,...,N and $\Delta t = 1$. (b) FFT squared modulus of y_n (complex conjugate of the Fourier coefficients, which determines the power), with $\Delta f = 1/(N\Delta t) = 1/512$, showing two closely spaced frequencies: 0.050 at n = 26.6, and 0.055 and n = 29.16 in the positive frequencies and –0.050 at n = 487.4 and –0.055 at n = 484.84 in the negative frequencies. (c) Close-up of the two positive frequencies. (d) Close-up of the two negative frequencies. The true frequencies (0.050 and 0.055; dashed red lines) do not coincide with the sampled frequencies (black dots) and do not recover the true squared modulus (i.e., true value of 512) of either of the frequencies.

length of the series, and the highest measurable frequency, $1/2(\Delta t) = f_{nyq}$, is the Nyquist frequency. The FFT stores the "Fourier coefficients" of the sines and cosines as a complex variable that is a function of frequency with a spacing, called the Rayleigh spacing, of $\Delta f = 1/N\Delta t$, for sampled frequencies $f_n = (n-1)\Delta f$, n = 1, ..., N. In practice, this FFT "mesh" resolves frequencies from 0 to $f_{nyquist} = 1/(2\Delta t)$, which is up to n = N/2+1 (for even N) and up to (N + 1)/2 (for odd N).

The Rayleigh spacing determines the spacing between spectral estimates along the frequency axis of the FFT, and sets the frequency resolution of the FFT. Synonyms for Rayleigh spacing include the terms "elementary bandwidth," "frequency bins," and "sampling frequency." The remaining sampled frequencies from $N/2 + 2 \rightarrow N$ map as negative frequencies from $N/2 \rightarrow 0$; real-valued time series will produce identical Fourier coefficients as the positive frequencies (in reverse order) (Figure 4.8). In practice, only the first N/2 Fourier coefficients are considered for applications involving real-valued data. For noise modeling (Section 4.3.6), it is important to remember that the first N/2 Fourier coefficients represent only half the variance of the input time series.

4.3.5.2   Direct Spectral Estimators, Leakage, and Tapers

The squared modulus of the FFT in Figure 4.8b is a direct spectral estimator and is known as the "Schuster," "classical," or "unsmoothed" periodogram:

$$S_D(f) = \left| \sum_{n=0}^{N-1} x(n)D(n)e^{-i2\pi nf} \right|^2$$

The exponential is shorthand for writing cosine and sine from Euler's formula: $e^{iy} = \cos y + i \sin y$. D(n) is a data window, also known as a "lag window," which multiplies the data time series x(n), i.e., which restricts it to a finite length. Technically speaking, the unsmoothed periodogram is multiplied by a rectangular ("Dirichlet") data window defined by the value of 1 over the time interval occupied by the time series and 0 elsewhere. The unsmoothed periodogram does not estimate the spectrum of the time series in Figure 4.8a very accurately: the time series has two sinusoids with equal amplitudes, but the FFT mesh imposed by evaluating the Fourier transform at the $\Delta f = 1/512$ spacing does not coincide with either frequency, and power is leaked into neighboring frequency bins.

Multiplication of the data window D(n) with the time series x(n) is equivalent to convolving the Fourier transform of D(n) with that of the time series. The Fourier transform of the Dirichlet window, called the Dirichlet "spectral window," is the sinc function, consisting of a "central lobe" of width Δf surrounded by an infinite series of sidelobes with decreasing power. This "ringing" is caused by the corners at the start and end of the Dirichlet window. The net result of convolving the sinc function with spectral peaks of the Fourier-transformed (infinite) time series is power leakage of approximately 10% (Table 1 of Durrani and Nightingale (1972) into the sidelobes).

To reduce this leakage, tapered data windows have been developed to smooth the window corners (Harris 1978). In Figure 4.9, the Dirichlet spectral window is compared with the spectral windows of the well-known Bartlett and Hann **tapers**. The "central lobes" of the spectral windows appear as half lobes centered on $f = 0$, and define the frequency resolution of the spectral estimator. The Dirichlet spectral window has the narrowest central lobe; the Bartlett and Hann central lobes are wider. In practice, central lobe width is defined in terms of "equivalent noise bandwidth" with $1.0\Delta f$ (Dirichlet), $1.333\Delta f$ (Bartlett), and $1.5\Delta f$ (Hann).

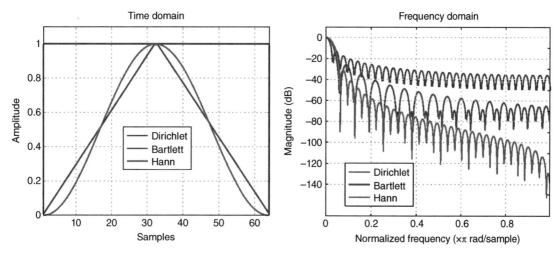

Figure 4.9 Three common data windows or "tapers" used in spectral analysis (left) and their Fourier transforms (right). The Dirichlet (also known as the "rectangular" or "boxcar") taper admits the entire time series from start to finish, whereas the other two tapers strongly down-weight the beginning and ending segments of the time series. These plots were generated in MATLAB using the Window Visualization Tool (for L = 64, wvtool(rectwin(L)); wvtool(bartlett(L)); and wvtool(hann(L)). Bartlett-tapered and Hann-tapered spectral estimates need to be corrected by a factor of 2 ("COHERENT GAIN" column in Table 1 of Harris (1978)). The Dirichlet Fourier transform has a central lobe with a half-power width of 0.89Δf; the Bartlett and Hann central lobes are broader at 1.28Δf and 1.44Δf, respectively (Table 1 of Harris (1978), "3.0-dB BW (BINS)" column). The "roll-off" of power outside of the center lobe is –6 dB per octave with the first sidelobe at –13 dB for the Dirichlet window, –12 dB per octave with the first sidelobe at –27 dB for the Bartlett window, and –18 dB per octave with the first sidelobe at –32 dB for the Hann window. (Note: dB (decibels) = $10 \log (P_2/P_1)$ where P_2 is power with respect to a reference power P_1.)

The Dirichlet window produces a spectral estimator with the narrowest frequency resolution, but at the cost of power leakage into the sidelobes. Leakage occurs when a true frequency component of the data time series does not coincide with a frequency bin, as shown in Figure 4.10; the Bartlett and Hann tapers somewhat restore leaked power back into the central lobe.

4.3.5.3 Spectral Estimator Statistics

The true strength of the spectral estimator lies in its statistics, which allows the practitioner to interpret the data spectrum. The probability distribution of the periodogram for Gaussian (normally)-distributed random variable was originally described as a Rayleigh distribution (Schuster 1897). This work laid the foundation for calculating the significance of estimated spectra in terms of the more general χ_n^2 distribution (Figure 4.11). The χ_n^2 distribution also provides an empirical means to determine the degrees of freedom (dof) n from the ratio of the mean and variance of the estimated periodogram (page 22 in Blackman and Tukey (1958); page 253 in Jenkins and Watts (1968); further information on dof bias in Elgar (1987)).

The unsmoothed (Dirichlet-windowed) periodogram is comprised of sine and cosine functions with coefficients contributing one dof each to the spectral estimate of a given frequency. In the FFT, the "Fourier coefficients" are stored in a complex variable. Thus, the unsmoothed periodogram has a χ_2^2 distribution; the n = 2 dof can be confirmed empirically using *estdof.m*

Figure 4.10 Effect of tapering on the same test time series as in Figure 4.9, here sampled at $\Delta t = 1$ for $N = 2048$ with two sinusoids at frequencies $f_1 = 0.050$ and $f_2 = 0.055$. (a) The Dirichlet-tapered test time series (top), Bartlett-tapered time series (middle), and Hann-tapered time series (bottom). (b) The periodograms of each tapered series is shown; arrows show attenuation of leakage by the Bartlett- and Hann-tapered periodograms. The inset shows that the Bartlett and Hann periodograms have recovered some of the power leaked by the Dirichlet periodogram, although they also fail to bring the power to the true modulus value of 2048. The leakage occurs because the frequencies do not coincide with the frequency bins proscribed by the FFT, although, leakage is substantially less here, even for the Dirichlet periodogram, than for the example shown in Figure 4.8, where the test time series was much shorter, at 512 points long.

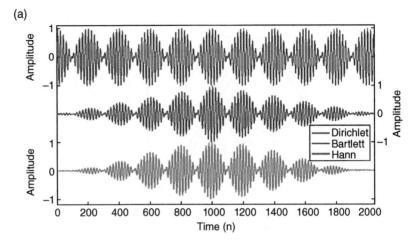

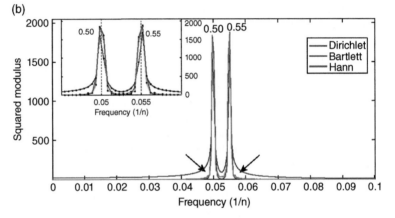

Figure 4.11 The χ_n^2 distribution reported as $f_n(x)$ versus x, for $n = 1$–20, calculated with *chisquare.m* (Appendix). $\{100(\alpha/2)\}\% = 95\%$ CLs are listed for selected n, so that the probability $P[\chi_n^2 \leq \text{lower CL}] = P[\chi_n^2 \geq \text{upper CL}] = \alpha/2$. These values are used in ratio with n, and multiplied with the spectral estimates to obtain upper and lower 95% CL constraints (see example in text). The definition of CL constraints for spectral estimates is required for hypothesis testing (Section 4.3.6).

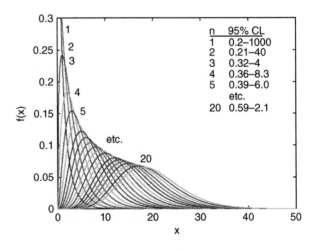

n	95% CL
1	0.2–1000
2	0.21–40
3	0.32–4
4	0.36–8.3
5	0.39–6.0
	etc.
20	0.59–2.1

(Appendix). The dof of the Bartlett or Hann-tapered periodogram are less straightforward owing to the loss of information from the tapering in the time domain, but gain in information from a wider central lobe (and lower sidelobes) in the frequency domain. Numerous books and papers report the "effective" dof's of smoothed periodograms for a variety of data windows (tapers), including the three discussed here (e.g., Table 6.6 in Jenkins and Watts (1968), Table 6.2 in Priestley (1981), Table 269 in Percival and Walden (1993)). For our three data windows, the effective dof per Fourier coefficient are Dirichlet ($n = 1$), Bartlett ($n = 3$), and Hann ($n = 8/3 = 2.667$).

Thus, for example, the Dirichlet-windowed (unsmoothed) periodogram, with $n = 2$, $S_D(f)$ has a 95% chance of falling within interval defined by the so-called lower and upper 95% confidence limits (CL): [$n \cdot S_D(f)$/upper CL, $n \cdot S_D(f)$/lower CL] = [$0.002 \cdot S_D(f)$, $9.5238 \cdot S_D(f)$]; the Bartlett-windowed periodogram, with $n = 2(3) = 6$, gives a 95% CL interval of [$6 \cdot S_D(f)/14.4$, $6 \cdot S_D(f)/1.2373$] = [$0.41667 \cdot S_D(f)$, $4.8493 \cdot S_D(f)$].

These low dof do not translate into reasonable constraints for spectrum estimates. However, averaging over adjacent frequencies (also known as "frequency merging") increases the dof and decreases the CL interval. At the same time, averaging reduces frequency resolution. Averaging can be carried out in the frequency domain, but it is usually done in the time domain by dividing the data time series into segments, either end-to-end and independent ("Bartlett's method," using the Dirichlet data window; Bartlett (1948, 1950)) or overlapping ("Welch-Overlapped-Segment-Averaging or WOSA method"; Welch (1967); now adapted for other data windows, for example, Schulz and Stategger (1997)), and subsequently averaging the segment periodograms in the frequency domain.

In the WOSA method, the data time series of length N is divided into n_{50} segments each of length N_{seg} that overlap each other by 50%, i.e., $n_{50} = 2N/N_{seg} - 1$. The effective dof of WOSA estimators depend further on the data window that is applied: $n_{eff} = n_{50}/(1 + 2c_{50}^2 - 2c_{50}^2/n_{50})$, where c_{50} is the 50% overlap correlation (e.g., 50% for the Dirichlet window, 25% for the Bartlett window, and 16.7% for the Hann window, see Table 1, rightmost column, of Harris (1978)). WOSA applied to the test time series (Figure 4.10) for different segment lengths shows improved leakage control and CL, although much of the improvement occurs with application of the Hann window, with diminishing returns for increased averaging, which broadens the spectral bandwidth, seriously encroaching the definition of the two frequencies (Figure 4.12).

4.3.5.4 Zero Padding

A common practice in spectrum estimation is zero padding. Many practitioners are unaware that it is being applied to their data, when, for example, the FFT used in the spectral estimation requires that the data time series have a length that is a power of 2, for example, in the SSA-MTM Toolkit (Ghil et al. 2002), and the default settings of MATLAB's FFT. The other motivation for zero padding is to interpolate the FFT mesh of spectrum estimates, as illustrated in Figure 4.13. Zero padding can also be used to estimate uncertainties on spectral peak frequency identification (e.g., Abe & Smith 2004).

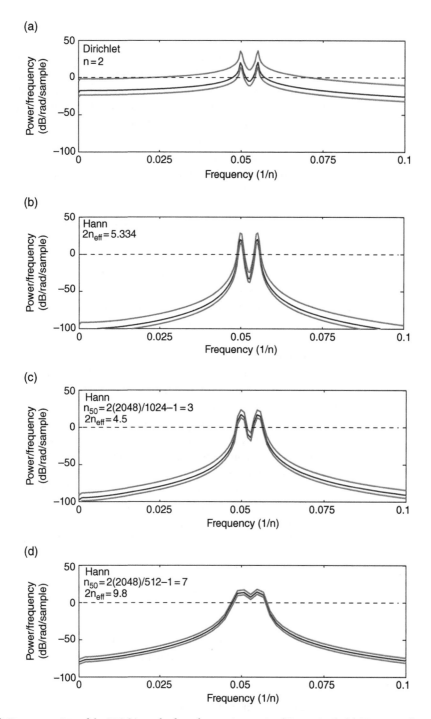

Figure 4.12 Demonstration of the WOSA method on the test time series (Figure 4.10). (a) Unaveraged Dirichlet periodogram with n = 2 dof from *spectrum.periodogram.m*. (b) Unaveraged Hann periodogram from *spectrum. periodogram.m*. (c) WOSA-averaged Hann periodogram from *spectrum.welch.m*, Nseg = 1024; d. WOSA-averaged Hann periodogram from *spectrum.welch.m*, Nseg = 512. The effective dof of the periodograms in (b–d) are given as $2n_{eff}$. Normalized frequency output from MATLAB was multiplied by 2 to convert to the frequency scale as in Figure 4.10. The gray lines are the lower and upper 95% CL of the spectral estimates. As dof increase, the CL becomes more equally distributed about the spectral estimates.

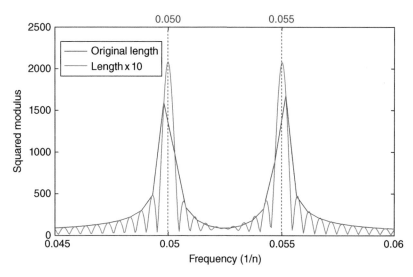

Figure 4.13 Benefit of zero padding is demonstrated for the test time series in Figure 4.10a. The blue curve is the Dirichlet periodogram of the time series, which is N = 2048 in length (same as in Figure 4.10b (inset) and Figure 4.12a), with an FFT mesh defined by $\Delta f = 1/2048$, i.e., no frequency bin (multiple of Δf) coincides with either of the two signal frequencies (0.050 and 0.055, see vertical dashed red lines). The green curve is the Dirichlet periodogram of the same test time series, padded with 0s to 10 times its original length (with zeros appended to the end of the time series prior to Fourier transformation). Now the FFT mesh is defined by $\Delta f = 1/20480$, with frequency bins very close to the true signal frequencies. The measured power is now equal between the two frequencies, reflecting their equal amplitude contributions in the time series. The Dirichlet "sidelobes" of this interpolated spectrum reveal the original frequency resolution of the time series.

Importantly, the (prewhitened) time series must be tapered prior to zero padding, so as not to institute a new source of spectral leakage from sudden transition to zeros at the end of the time series. In the literature, spectra reported with high-frequency "multilobed" features, such as those in Figure 4.13, or embellishing the tops of smoothed spectral peaks may have been zero padded, and should not be interpreted or used to evaluate the outcome of hypothesis tests (Section 4.3.6).

4.3.5.5 Indirect Spectral Estimators and the BT Correlogram

The autocorrelation function $\rho(n)$ of a time series equals the squared modulus of the Fourier transform of the time series (the Wiener–Khinchin theorem; Blackman and Tukey (1958)). This relationship leads to an alternative estimate of the power spectrum:

$$S_D^{BT}(f) = \left| \sum_{n=-(M-1)}^{M-1} \rho(n) D(n) e^{-i2\pi nf} \right|^2$$

where $M \leq N$ equals the number of autocorrelation "lag" coefficients used in the estimates and sets the smoothing factor of the estimates. This is the BT correlogram, also known as the "lag window spectral estimator" or "indirect

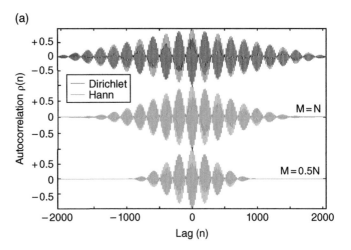

(a)

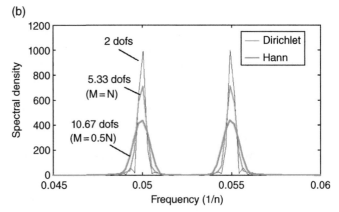

(b)

Figure 4.14 Autocorrelation and the BT correlogram of the test time series from Figure 4.10. (a) Top: Autocorrelation function of the time series computed from xcorr.m (xcorr(signal) and normalized); middle: autocorrelation function multiplied by a Hann lag window of length M = N; bottom: autocorrelation function multiplied by a Hann lag window of length M = 0.5 N. (b) BR correlogram estimates of the time series displayed in a; effective dof are shown.

spectral estimator". D(n) is the data "lag" window applied analogously as in the direct spectral estimator discussed above. The effective dof are larger for smaller M, and defined in a straightforward way for windowed BT estimates, per Fourier coefficient: N/M for the Dirichlet lag window, 3N/M for the Bartlett lag window, and 8N/3N for the Hann lag window (e.g., Table 6.2 in Priestley (1981)).

BT correlogram estimation is illustrated in Figure 4.14 for the Dirichlet and two Hann lag windows. The BT correlogram and direct spectral estimator are not interchangeable: the autocorrelation function is naturally "tapered" (owing to the limited length of the time series) and is twice as long ($N_\rho = 2 \times N - 1 = 4095$), which results in a finer-scale FFT mesh. Thomson (1977, 1990, 2009) warns that the BT correlogram can produce extremely biased spectral estimates of some time series and advises against its use.

4.3.5.6 Thomson Multitapers

Over the past 50 years, dozens of single tapers have been designed to optimize different aspects of spectrum estimation (e.g., Harris 1978). The objective of all of these tapers is to control spectral leakage from the center lobe, increase the statistical stability (maximize the dof), minimize bias (due to the preferential sampling of the center part of the time series), and retain the highest (narrowest) possible frequency resolution. The quest to satisfy all of these criteria led to the concept of "multitapers." Families of functions known as "discrete prolate spheroidal sequences (DPSS)," each function independent from the others, and when summed together approximate a Dirichlet data window (Slepian 1978). DPSS are also known as "Slepian sequences" in honor of the mathematician David Slepian (1923–2007). If the Slepian sequences in a family are individually applied as tapers to a time series, and each Slepian-tapered series is then Fourier transformed, the set of Fourier transforms may be averaged together to produce a smoothed spectral estimator, with each transform nominally contributing 2 dof (Thomson 1982):

$$S_{x,D}^K(f) = \frac{1}{(K+1)} \sum_{k=0}^{K-1} \left| X_D^K(f) \right|^2 \text{ and } X_D^K(f) = \sum_{n=0}^{N-1} x(n) D_k(n) e^{-i2\pi n f}$$

where $x(n)$ is the time series and $D_k(n)$ are Slepian sequences. The Slepian sequences maximize variance (power) within a narrow band of frequencies W spanning several Δf with respect to the total resolvable band of frequencies in a time series of length N, i.e., $\pm f_{Nyq}$. The Slepian sequences are solved as eigenvectors with associated eigenvalues indicating the bias of each eigenvector. The eigenvectors with eigenvalues closest to 1 are retained as the "minimum-bias" set of "eigentapers" to be applied to the time series. The time-bandwidth product is selected as $P = NW$, and the maximization problem is solved (e.g., Park et al. 1987) to yield a set of $P\pi$ prolate eigentapers (usually shortened to "tapers"), of which the lowest orders 0 through K-1, where K = 2P, have eigenvalues very close to 1. The Fourier transform of an "eigentapered" time series is an "eigenspectrum." Usually the eigentaper of order K – 1 has an eigenvalue that is significantly lower than 1 and is dropped from the set. For cyclostratigraphy, P usually ranges from 2 to 6 and can be fractional (e.g., 3.3 and 4.8). The Fourier transforms of the individual tapers from a single family reveal that as taper order increases, the central lobe of width W is sampled increasingly toward the outer edge of W (e.g., Figure 2 in Thomson (1982); Figure 3 in Park et al. (1987)). In the time domain, each taper in a family samples a different part of the time series (Figure 4.15a–d).

Multitaper spectral estimates of the test time series of length N = 2048 are compared with the unsmoothed (Dirichlet) periodogram (Figure 4.15e). For example, "5π," refers to the averaging bandwidth imposed by the application of a set of nine 5π Slepian tapers (i.e., orders 0–8), where the value P = 5 has been selected by the practitioner. The "5" corresponds to an averaging bandwidth W = 5/2048 = 0.0024414. The bandwidth W also sets the dof listed next to the displayed spectra on the right, nominally as 2NW (with further discussion in Section 4.3.5.7). All of the displayed multitaper estimates have high consistency (many dof), and the 2π estimator in particular shows high definition of the two signal sinusoids, with 6 dof at only twice the frequency resolution compared to the 2 dof and Δf resolution of the unsmoothed (Dirichlet) periodogram. Figure 4.16 shows the 95% CL of the 2π multitaper spectrum in the same

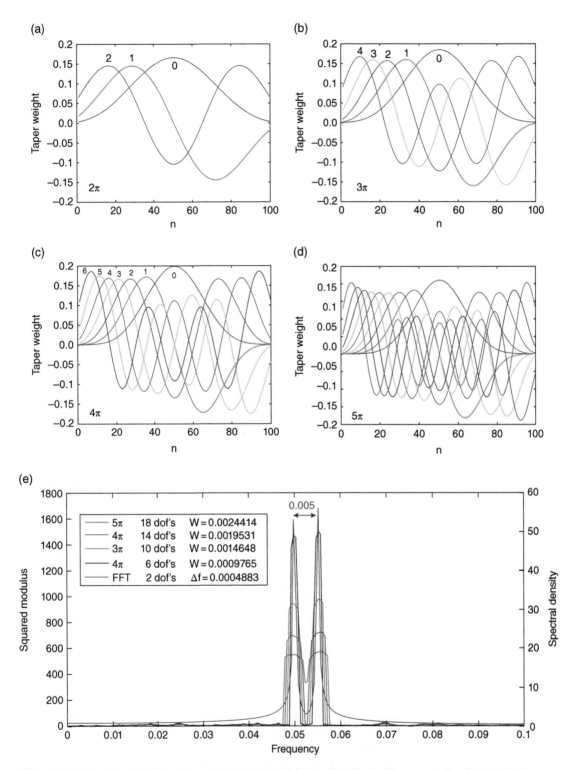

Figure 4.15 Examples of multitaper spectral estimators. Multitaper families (a–d) were calculated in MATLAB using *dpss.m*, e.g., [E,V] = dpss(100,2) for 2π multitapers defined for a length of 100 points. The "E's" are the displayed eigentapers, and the "V's" are the associated eigenvalues (not shown). Numbers indicate order. The tapers may be rescaled to the length N and sample rate Δt of the time series that is to be tapered. The sum of the absolute values of the tapers in a given family approximates a boxcar (Dirichlet) window. (e) [p,w] = pmtm(signal,2) for the 2π multitaper power spectrum of the time series in Figure 4.10, and converting radial frequency w to linear frequency, f = w/(2πΔt), is displayed for each of the multitaper estimators; the unsmoothed (Dirichlet) periodogram, designated as FFT, is shown for comparison.

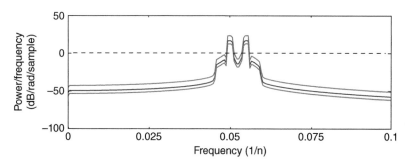

Figure 4.16 2π MTM estimated spectrum of the test time series of Figure 4.10, using *spectrum.mtm.m*, with the lower and upper 95% CL shown for 6 dof (see also Figure 4.15 for display with linear y-scale). While there are four tapers that qualify for use (and potentially $2 \times 4 = 8$ dof), MATLAB drops the last taper owing to its relatively low eigenvalue (0.7219). These results may be compared with the 95% CL of single-tapered estimates displayed in Figure 4.12.

units as the single-taper estimates in Figure 4.12. This example demonstrates the strength of the multitaper estimator: the adjustable bandwidth W solves many resolution problems while maintaining the high statistical stability (multiple dof). Thomson (1982) also describes a "harmonic line" F-test that examines the spectrum at the resolution of Δf for the presence of significant sinusoids (see Section 4.3.6.2).

4.3.5.7 Adaptive Weighting

In natural data spectra, power is usually unevenly distributed with frequency. Frequencies with high power generally (but not always) host more information than bands with low power. Moreover, high-order eigentapers tend to contribute more bias that preferentially affects the low power spectral estimates. To counter these problems, a "data adaptive weighting" strategy was developed to down-weight low power spectral estimates based on the data spectrum and bias from the kth eigentaper (Section V in Thomson (1982)). The resultant adaptive weights, $d_k(f)$, are unique to the input data time series and eigentapers being used in the spectral estimation. The weights are applied to the eigenspectra prior to the final averaging. The $d_k(f)$ also provide an empirical estimate of the effective dof as a function of frequency, assigning fewer dof to spectral regions with lower power:

$$\upsilon(f) = 2\sum_{k=0}^{K-1}\left|d_k(f)\right|^2$$

This contrasts with mathematical statistics that provide a single estimate of dof to be applied across the entire spectrum, for single tapered estimates such as those in Figure 4.12, as well as for unweighted multitapered spectra such as in Figure 4.16. The effect on adaptive weighting and $\upsilon(f)$ from different distributions of power is illustrated in Figure 4.17. In these examples, signal is deeply embedded in noise. The noiseless signal has $\upsilon(f)$ with the maximum number of 14 dof (i.e., 7 eigenspectra contribute 2 dof each) in the bands of width W surrounding the two signal frequencies; elsewhere $\upsilon(f)$ is at a minimum of 2. The power spectra of the two noisy time series just resolve the two signal frequencies; $\upsilon(f)$ tracks the estimated power level,

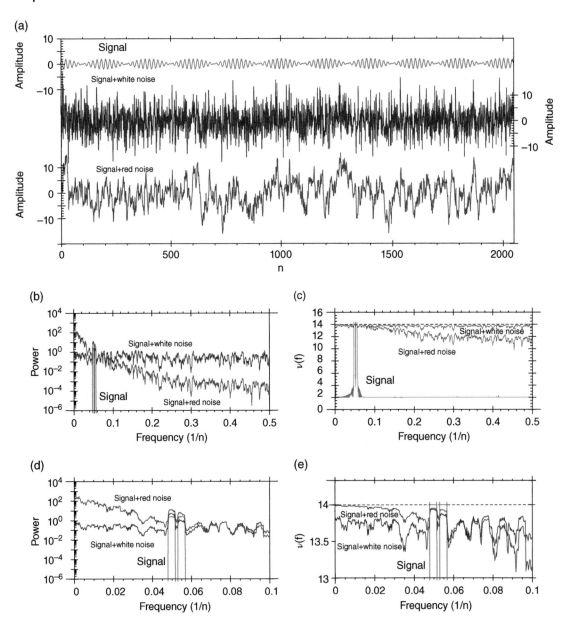

Figure 4.17 Effective degrees of freedom ν(f) estimated for three time series using seven 4π multitapers. (a) Top: noiseless signal of length N = 2048 represented by the test time series of Figure 4.10a (variance = 1); middle: same signal plus white Gaussian noise (variance = 25), computed using *randn.m* (5× randn(2048,1)); bottom: same signal plus strong AR(1) red noise (variance = 25), computed by submitting the same Gaussian white noise used just previously into *markovseries.m* and setting ρ = 0.9 (details in Section 4.3.6) and rescaling the variance to 25. (b) Adaptive weighted 4π multitaper spectra of the three time series shown in (a). (c) Degrees of freedom ν(f) estimated from the adaptive weights of the three spectra in (b). (d) Detail of (b) centered on the two signal frequencies at f=0.050 and f=0.055. (e) Detail of (c) centered on the two signal frequencies at f = 0.050 and f = 0.055. The horizontal dashed line in (c) and (e) represents the dof that are assigned (14 in this case) when adaptive weighting is not applied.

with local maxima at the signal frequencies, but never reaching the maximum of 14 dof. SSA-MTM Toolkit (Ghil et al. 2002) applies adaptive weights to the power spectrum, but does not estimate $\upsilon(f)$ for hypothesis testing, instead using only maximum dof (e.g., the horizontal dashed line at 14 in Figure 4.17c). *Analyseries* (Paillard et al. 1996) does not calculate adaptive weights and reports unweighted MTM power spectra, known as "high-resolution" spectra (Thomson 1982).

4.3.6   *Hypothesis Testing and Noise Modeling*

The usual question asked of cyclostratigraphy is whether or not a signal with Milankovitch cycles is present. The power spectrum is consulted to assess narrow frequency bands of elevated power with nonrandom variance at the Milankovitch frequencies ("signal"). In natural data, uncorrelated, random variance ("noise") is also present and must be distinguished from the signal. Noise tends to occur at all frequencies, also known as the "continuum," and has a power level that is generally lower than that of the signal. The simulated time series in Figure 4.17 are typical of signal + noise combinations found in nature; in fact, in these two series, the noise level is quite high and challenging to distinguish from the embedded signal, as demonstrated below.

The signal versus noise problem in spectral analysis is approached through modeling and hypothesis testing. Signal is modeled as a process that is characterized by a frequency, or set of frequencies, of variation with persistent magnitude and constant phase. Noise is characterized by variation with contributions at all resolvable frequencies, and serves as the "null model" in the following hypothesis testing procedure:

1   The "null hypothesis" H_0 is that the data results from a random process, here represented by the noise model (see the following sections for commonly used models).
2   The "alternative hypothesis" H_A is that the data represents a combination of a nonrandom process and random chance.
3   A "test statistic" is used to assess the validity of H_0. In this case, the test statistic is the power spectrum.
4   The probability P is computed for the test statistic to evaluate whether it is at least as significant as the case for a true H_0. Smaller P is stronger evidence against H_0.
5   Compare P to a significance value α, e.g., 0.05; if $P \le \alpha$, H_0 is ruled out and H_A is accepted.

4.3.6.1   Spectral Noise: The Null Model

Two statistical models of noise are in wide use today: autoregressive (Markovian) noise and power law ($1/f^\alpha$) noise. These models recognize that noise spectra can have different frequency distributions depending on the process under consideration and its measurement. White noise has equal contributions of power at all frequencies; the white noise spectrum is a simple horizontal line. However many processes that are noise-like exhibit "memory," in that recent behavior influences the current behavior of the process. This memory suppresses high frequencies and amplifies low frequencies, which "reddens" the spectrum.

Autoregressive noise is modeled as a first-order linear Markov process (Gilman et al. 1963):

$$r(n) = \rho r(n-1) + e(n)$$

where r is the "autocorrelated" random process, ρ $(0 \le \rho < 1)$ relates the previous $(n-1)$ to the current (n) observation of r, and e is the Gaussian white noise. This is also known as a first-order autoregressive AR(1) model (Jenkins & Watts 1968). Direct inversion of the above equation reveals that ρ is the lag–1 autocorrelation coefficient of r (Priestley 1981). The power spectrum of r is:

$$R(f) = R_0 \frac{1-\rho^2}{1-\rho\cos(\pi f/f_{Nyquist}) + \rho^2}$$

where $R_0 = \sigma_r^2/(1-\rho^2)$ is the average power of the spectrum, with σ_r^2 the variance of r. R(f) is shown in Figure 4.18 for different ρ. When $\rho = 0$, the model reduces to Gaussian white noise with power level R_0; as ρ increases, power is progressively shifted into the low frequencies at the expense of power in the high frequencies. A realization of AR(1) noise with $\rho = 0.9$ is shown in Figure 4.17b, where it is displayed as $\log_{10}$(power) vs. linear frequency combined with the test time series with the two frequencies (signal).

Power law noise, or "1/f^a" noise, is characterized by reddened spectra with specific slopes in $\log_{10}$(power) versus $\log_{10}$(frequency) space. White noise is characterized by a=0. Flicker noise is characterized by a=1 and is observed widely in both natural and artificial phenomena (Press 1978; Milotti 2002) and is a "balance between randomness and correlation at all timescales" (Voss 1979). While it is the most commonly observed noise, it is the least understood. Brownian noise is characterized by a=2, also known as "random walk" noise or a Wiener stochastic process, with characteristics of the well-known Brownian motion (Gillispie 1996). The spectra of these two power law models (Figure 4.18) show that this noise is distinctively different from the AR(1) noise models, with comparatively greater inflation of power in the lowest frequencies. Recently, autoregressive model approximations have been developed for 1/f^a noise, which may lead to new null models.

Hypothesis testing of data spectra has traditionally relied on the autoregressive noise for the null model, with its readily adaptable parameterization, i.e., use of the lag–1 autocorrelation coefficient of the data time series for computing R(f). An example of a hypothesis test of the "signal + red noise" time series using AR(1) as the null model is shown in Figure 4.19. Only two spectral bands exceed the 99% CL, at the signal frequencies, 0.050 and 0.055.

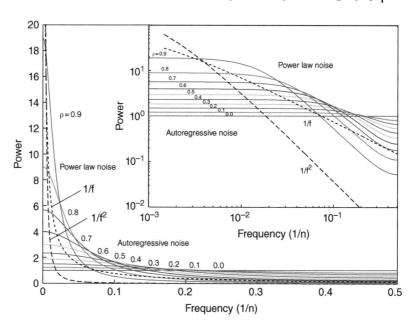

Figure 4.18 Spectral noise models. The main plot shows autoregressive (Markovian) spectral noise models (computed with *arnoisemodel.m* for different values for ρ) and power law noise models for 1/f and 1/f² (computed with *alphanoise.m* and rescaling the variance to 1, i.e., that of the autoregressive noise models), with 1:1 and 1:2 slopes in $\log_{10}$(power) vs. $\log_{10}$(frequency), as depicted in the inset.

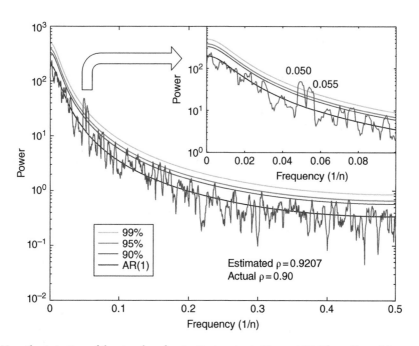

Figure 4.19 Hypothesis testing of the signal+red noise time series in Figure 4.17. The null model was calculated as an AR(1) model using Husson's *reconf.m*. The 4π MTM adaptive weighted spectrum (red curve) was assigned a constant 14 dof for all frequencies (dashed horizontal line in Figure 4.17), from which CLs were estimated at 85, 90, 95, and 99% levels. Instead of plotting each of these CLs with respect to the spectrum, the upper CL values were applied instead to the AR(1) model, which on its own has 1000s of dof (thus has CLs that are not appreciably different from the model estimate). This display is the convention that is used in the literature.

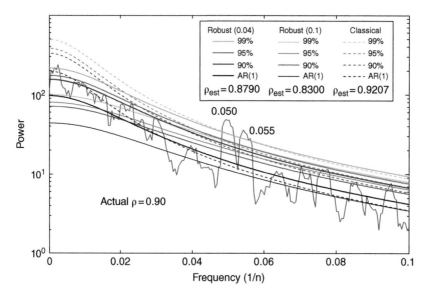

Figure 4.20 Classical and robust red noise modeling of the "signal+red noise" time series from Figure 4.17. The actual red noise in the time series has $\rho = 0.90$, but the classical red noise model overestimates $\rho = 0.9207$. The robust red noise models calculated using the SSA-MTM Toolkit with 0.1 and 0.04 median filter lengths and linear fitting result in underestimated $\rho = 0.8300$ and $\rho = 0.8790$. Alternative log fitting (not shown) results in $\rho = 0.8600$ for a median filter of length 0.1, and $\rho = 0.9100$ for a shorter median filter of length 0.04, which reduces the influence of the two signal frequencies on the modeled noise for $f \leq 0.04$.

But also of significance is that the estimated $\rho = 0.9207$ from the lag – 1 of the signal + red noise autocorrelation function is larger than the actual red noise component of the time series ($\rho = 0.9$). Thus, the AR(1) model, also known as the "classical red noise model," is biased: R(f) overestimates noise at low frequencies, and to a lesser degree, underestimates noise at high frequencies.

The bias from estimating the AR(1) null model from a time series containing both signal and noise was recognized by Mann and Lees (1996) who introduced "robust red noise" estimation. This procedure removes narrow band spectral components prior to AR(1) modeling on the premise that such components represent signal. Two methods were used to identify narrow band signal components: a harmonic F-test to detect individual lines (see Section 4.3.6.2) and median smoothing of the data spectrum to reject outliers presumed to represent signal.

Comparison of "classical" versus "robust" red noise modeling (Figure 4.20) shows that the former overestimates the noise, but the latter has adjustable parameters (median filter length, and linear vs. log fitting) that can be searched for the ρ closest to true ρ. Of course in practical applications, true ρ is unknown. Thus, a procedure is needed to identify the most correctly estimated ρ which has yet to be developed. Meyers (2012) has described an alternative "lowess" smoothing of the spectrum in robust AR(1) modeling to optimize ρ estimation. These classical and robust AR(1) models, despite the shortcomings, are the conventional null models presently used in cyclostratigraphy. Figure 4.21 demonstrates application of both models on the Arguis

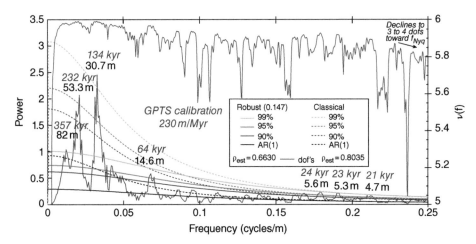

Figure 4.21 2ð multitaper spectral estimation of the Arguis ARM stratigraphic series depicted in Figure 4.2, here linearly interpolated to a constant Δd = 0.6757 m, with a 10% weighted (lowess) average removed (80 m averaging window, see Figure 4.4). The average sample rate for the original series is 0.6757 ± 0.58 m, which converts to an average $f_{Nyq} = 1/(2 \times 0.6757\,m) = 0.74$ cycles/m. The spectrum is adaptive weighted; the effective dof are computed from the same adaptive weights, and are stable, fluctuating slightly around 5.8. For convenience, the CLs of both models were estimated using 6 dof (the small fluctuations in dof produce very small changes in the CLs). The classical red noise model was computed using Husson's *redconf.m*; the robust red noise model was computed using the SSA-MTM Toolkit with a 0.147 cycle/m median filter with linear fitting.

ARM series, that has been resampled to a uniform 0.6757 m spacing, i.e., the average spacing for the entire series, necessary for accurate calculation of the lag–1 autocorrelation coefficient. The effective dof over f = [0,0.1 cycles/m] averages about 5.8; for convenience 6 dof were assumed for the CLs of both models. The two noise (null) models estimated from the Arguis ARM series are quite different from one another compared with the test series case in Figure 4.20. Spectral power is concentrated in the very low frequencies, which greatly influences the classical red noise model which estimates ρ = 0.8035. The robust model rejects spectral peak outliers as noise, i.e., the high power peaks in the low frequencies; this results in a substantially "less red" null model, with an estimated ρ = 0.6630. The labeled spectral peaks all exceed at least the 95% CL for either model, and based on chronological constraints from the GPTS (Figure 4.2) may be evidence for Milankovitch cycles. The dramatic difference between classical and robust estimates of the noise indicates that the latter has detected large spectral outliers in the low frequencies, i.e., the 30–80 m variations; this suggests that true ρ of the noise is closer to the robust estimated ρ = 0.6630. Sensitivity studies of the robust model that varies the median filter length, as demonstrated in Figure 4.20, may provide additional clues about true ρ.

The selection of the CL that is most appropriate for interpretation is an open issue. Statisticians and scientists who work with highly controlled signals adopt very strict CLs in excess of 99%. However, in cyclostratigraphy, variable accumulation rates result in poorly controlled signals and cause true lines and spectra to be recorded at different wavelengths, which

(a)

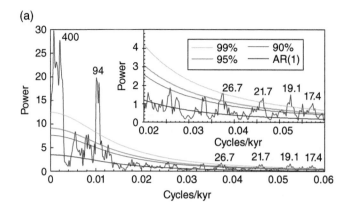

(b)

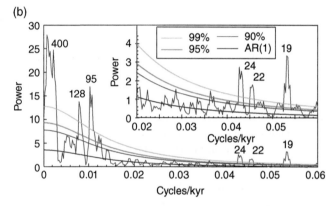

Figure 4.22 Robust red noise modeling of astronomically tuned Arguis ARM time series compared with 2π multitapered spectra. (a) The twice eccentricity-tuned series with estimated $\rho = 0.9070$ and (b) twice eccentricity then precession index-tuned series with estimated $\rho = 0.9190$ (Figure 5.5 in Chapter 5). All peaks exceeding the 99% CL are labeled in periodicity (kyr). The insets zoom in on the obliquity and precession index bands. In (a), despite tuning to the eccentricity two times, the 128 kyr component of the orbital eccentricity is not resolved; but subsequent precession index tuning resolves the 128 kyr component. The average Δt of both series is 3 kyr, or an average $f_{nyq} = 0.1667$ cycles/kyr. The spectra were computed using the SSA-MTM toolkit, with robust red noise parameters set for a 0.033 cycles/kyr median filter, linear fitting and 6 dof for all frequencies.

blurs their appearance in an averaged power spectrum. Therefore, in the stratigraphic domain, assuming an arbitrarily strict CL is ill advised. Instead, reporting multiple CLs, as has become customary, provides information about the relative statistical significance of spectral peaks. Later, for example upon astronomical tuning, high CLs can be demanded of the data (Figure 4.22). Accumulation rate effects can be preliminarily assessed using time-frequency techniques (Section 4.3.7).

4.3.6.2 Harmonic F-Test

Thomson (1982) developed a multitaper test for harmonic (spectral) lines, i.e., power that is consistently confined to single frequency bins in a time series, as for example, the two sinusoids in the test time series in Figure 4.17. The K eigencoefficients (the Fourier coefficients of each of the K eigentapered series) at frequency f are combined to estimate the mean modulus, or amplitude $\mu(f)$, for which $\mu(f)^2$ is a "signal" power estimate with 2 dof. Subtraction of $\mu(f)$ from the K individual eigencoefficients at frequency f is used to estimate a residual modulus $\mu_{res}(f)$, for which $\mu_{res}(f)^2$ is a "noise" power estimate with $\nu(f) - 2$ dof. The ratio, $F = \mu(f)^2/\mu_{res}(f)^2$, is a signal-to-noise ratio of variances with an F-distribution of 2 and $\nu(f) - 2$ dof. It is applied as a conventional F-test: when an F value exceeds the F-distribution at a specified $1 - \alpha$ level, a harmonic line may be inferred that is significantly above the level of the noise.

Figure 4.23 shows the results of the harmonic F-test of the "signal + red noise" test series, which is an extremely noisy time series with a signal with a variance of 1 (constrained to frequencies 0.050 and 0.055) and red noise ($\rho = 0.9$) with a variance of 25. The signal is actually not visually discernable in the time series (Figure 4.17a); it is notable that the spectral analysis techniques demonstrated thus far have been able to detect it. The reason for this is that over the entire length of the time series the signal maintains perfect repeatability with constant ("coherent") phase. The noise, on the other hand, has irregular phasing with variance at any given frequency that can "cancel out" when evaluated over the entire series length. The question is whether in the presence of the strong noise in this test time series, F-testing will be able to identify the signal frequencies as harmonic lines.

At very low frequencies ($f < 0.04$), the noise dominates the spectrum (Figures 4.23a and b); nonetheless the two signal frequencies are clearly distinguished. Over 90% of the power occurs from $f = [0,0.1]$, and the F-testing should therefore focus on this range. The dof used in the F-testing is based on $\nu(f) - 2$ (Figures 4.23c and d) where $\nu(f)$ was previously shown in Figure 4.17c. A useful by-product of the F-test procedure is an estimate of the amplitude spectrum (Figure 4.23e and f), which indicates that the red noise has added variance to both frequencies (amplitudes slightly over 1.0, the true amplitude). Several of the noise-only frequencies indicate possible coherent phase behavior, registering F-values with significance levels in excess of 99% (Figure 4.23j), but the highest significance levels occur at the two signal frequencies at 0.050 and 0.055. In practice, a spectrum with $N/2 + 1$ frequencies will register $\alpha(N/2 + 1)$ F-tests exceeding the $1 - \alpha$ significance level. Here, with $(4N)/2 + 1 = 4097$ frequencies, and $\alpha = 0.01$, we would expect ~41 F-tests at 0.99 and higher; 39 such F-tests are recorded in Figure 4.23i.

In summary, the harmonic F-test detected the two lines in the signal + red noise test time series. However, other high F-values (37 in total) generated by the noise produced numerous "false positives" that could have been (incorrectly) interpreted as lines. Our interpretation was possible based on clues from the power spectrum indicating elevated power at the line frequencies, and *a priori* knowledge of the existence of the lines. However, for time series with unknown spectral content, which is always the case in cyclostratigraphy, F-testing must be undertaken with great caution. Thomson (2009) advises

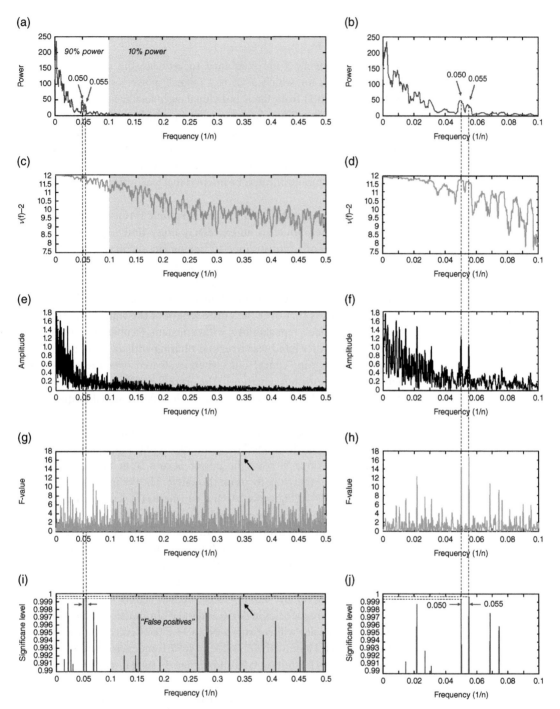

Figure 4.23 Harmonic F-testing of the signal+red noise test series with seven 4π multitapers using *pmtm,m, mtmdofs.m,* and *ftestmtm.m* (see Appendix). Vertical dashed lines indicate the positions of the two signal frequencies at 0.050 and 0.055 (red arrows). The shaded area of the frequency space indicates the region with 10% of the total power. (a) Power spectrum over $[0,f_{Nyq}]$; (b) power spectrum over $[0,0.1]$; (c) dof for the F-test denominator over $[0,f_{Nyq}]$; (d) dof for the F-test denominator over $[0,0.1]$; (e) amplitude spectrum over $[0,f_{Nyq}]$; (f) amplitude spectrum over $[0,0.1]$; (g) F-value (signal-to-noise ratio) over $[0,f_{Nyq}]$; (h) F-value (signal-to-noise ratio) over $[0,0.1]$; (i) significance level of F-values over $[0,f_{Nyq}]$; (j) significance level of F-values over $[0,0.1]$. Significance levels were computed by integrating the F-distribution of 2 and $\nu(f)$-2 degrees of freedom from zero to the estimated F-value at frequency (f). The high frequency region with only 10% of the total power is prone to "false positives" (i.e., as $\mu_{res}(f) \to 0$, F-value $\to \infty$.) Horizontal dashed lines in (i) and (j) indicate the significance levels of the two signal frequencies, which have the first and third highest levels over $[0,f_{Nyq}]$ and are first and second over $[0,0.1]$. The second highest F-test at f=0.34 (black arrow) is associated with the highest F-value, but $\nu(f)-2=9.9$ reduces its significance to overall second place. This analysis was computed with zero padding to 4N.

that F-tests should be interpreted in conjunction with power spectrum structure (as was done here) and provides additional notes on expected outcomes
of F-testing using cyclostratigraphy as an example.

4.3.7 Time-Frequency Analysis

Natural systems often "drift," and their frequencies and magnitudes can
change, slowly (e.g., groundwater flow), or suddenly (e.g., seismic events), or
may have quasi-periodic behavior (e.g., Milankovitch-forced climatic
change). Analytical methods that track time-frequency changes are needed
to assess such processes. Here, two representative methods are illustrated:
evolutionary spectrograms and complex signal analysis.

4.3.7.1 The Evolutionary Spectrogram

The evolutionary spectrogram can take on numerous forms and is the
simple application of a spectral estimator with a moving (or "running")
time window through power with respect to frequency on the x-axis and
time on the y-axis. 3D displays can also be effective (not shown here).
The spectrogram shown in Figure 4.24 displays a running "unsmoothed
periodogram" (Section 4.3.5) of the La2004 astronomical model of the
Earth's orbital eccentricity compared with the two astronomically tuned
versions of the Arguis ARM series (Figure 4.22; see also Chapter 5).
These spectrograms highlight the intricate quasi-periodicity of the
orbital eccentricity parameter, in which the ~100 kyr spectral terms
experience an interval of low power relative to the 405 kyr term in the
interval centered on 37.5 Ma. Experimentation with window length is an
important aspect of the spectrogram; if it is too long, high frequencies
will be smoothed out, and if it is too short, low frequencies will not be
measured adequately.

4.3.7.2 Complex Signal Analysis

Complex (or quadrature) signal analysis is a classical technique used to
estimate instantaneous amplitude, phase, and frequency attributes of a
real signal as a function of time (Taner et al. 1979). The complex repre-
sentation of a real signal g(t) is given as $G(t) = g(t) + ig^*(t)$, with $i^2 = -1$
and "*" denoting complex conjugate, where $g^*(t)$ is obtained by Hilbert
transformation of g(t).

The technique is demonstrated in Figure 4.25, in which a 1.47 kyr cycle
(Dansgaard-Oeschger scale) experiences a 40 kyr frequency modulation
and a 20 kyr amplitude modulation over a 100 kyr interval. The complex
signal analysis shows that the amplitude modulation imposes "singularities"
in f(t) whenever the amplitude passes through zero, also producing discon-
tinuities in the instantaneous phase. Such "singularities" can be picked up by
depositional hiatuses or other perturbations not related to amplitude forcing
and can also occur as the result of band-pass filtering.

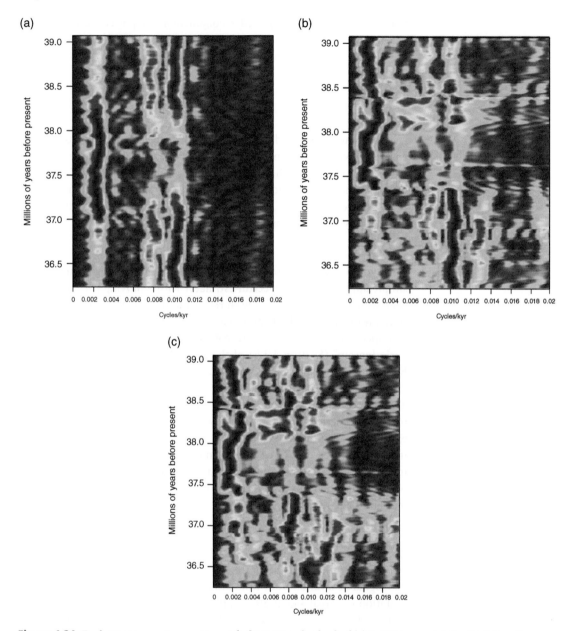

Figure 4.24 Evolutionary spectrograms over f = [0, 0.02 cycles/kyr] of (a) La2004 eccentricity, (b) 2-iteration eccentricity-tuned Arguis ARM series (see also power spectrum in Figure 4.22a), and (c) 2-iteration then precession index-tuned Arguis ARM time series (see also power spectrum in Figure 4.22b) from 36 to 40 Ma (details in Chapter 5). An 800 kyr window was used, stepping through the series at 5 kyr intervals, calculating the unsmoothed periodogram. Each periodogram is normalized to itself, which helps to track the evolution of the major spectral components of each window. The y-axis displays the first periodogram at 36 Ma and the last periodogram at 40 Ma – 0.8 myr = 9.2 Ma. Each contributing spectrum has been normalized to itself, in order to track the major contributing spectral components throughout the time series. These spectrograms were computed with *evofft.m* (see Appendix).

Figure 4.25 Complex signal analysis of the amplitude and frequency modulated time series g(t). This time series was calculated by sampling a sinusoid $h(t_n)$ of period 1/1.47 cycles/kyr at intervals $t_n = n\Delta t$, with $\Delta t = 4$ years, and $n = 1, 2, \ldots, 25{,}000$, and amplitude modulated by $A(t_n) = \sin[2\pi t_n/(20\,\text{kyr})]$, such that $g(t_n) = A(t_n) \cdot h(t_n)$. For the frequency modulation, the timescale t_n was replaced with $t'_n = t_n + \Delta t[1 + \sin(2\pi t_n/40\,\text{kyr})]$, to give $g(t'_n)$, which was then resampled to a uniform rate of $\Delta t = 4$ years as g(t). (a) The modulated time series g(t); (b) the complex signal G(t) obtained using *hilbertsignal.m* (see Appendix), shown in 3D, with the complex axis normal to the page; (c) instantaneous amplitude A(t); (d) instantaneous phase $\theta(t)$; and (e) instantaneous frequency f(t).

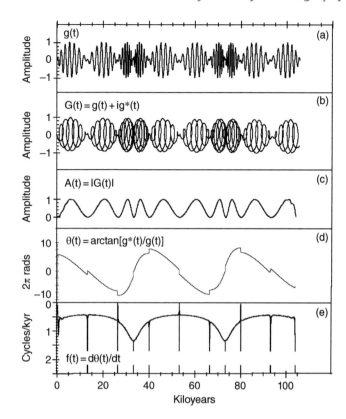

The input signal g(t) must have a zero mean and with a relatively narrow band; most applications are preceded by band-pass filtering, for example, filtering for the precession index, and analyzing the instantaneous amplitude of the band-pass filtered signal to seek evidence for an orbital eccentricity modulator (see Chapter 5).

This technique could be, but has not yet been, used to estimate accumulation rate changes from stratigraphic sequences experiencing variable cycle wavelengths. To whit, the example in Figure 4.25 was created to illustrate Milankovitch influences on oxygen isotope Dansgaard-Oeschger cycles in the GRIP (Greenland Icesheet Project) ice core; in particular, a strong 40 kyr (obliquity) frequency modulation was found to affect the record, most probably due to systematic error in the ice-flow based core chronology (Hinnov et al. 2002).

4.3.8 Coherency and Cross-Phase Analysis

Multivariate time series analysis is used to investigate two or more time series that are hypothesized to have a relationship, e.g., cause and effect. Cross-correlation analysis as a function of frequency is a fundamental tool for such comparisons, with clearly defined statistics and hypothesis testing capabilities. Using the Thomson multitaper approach, the cross-correlation of two time series x(n) and y(n) is assessed with the coherence and cross-phase functions:

$$C(f) = \frac{S_{x,y,D}^K(f)}{\sqrt{S_{x,D}^K(f)}\sqrt{S_{y,D}^K(f)}} \text{ and } f(f) = \arctan\left[\frac{\text{imag}(S_{x,y,D}^K(f))}{\text{real}(S_{x,y,D}^K(f))}\right]$$

where

$$S_{x,y,D}^K(f) = \frac{1}{(K+1)}\sum_{k=0}^{K-1}[S_{x,D}^K(f) * S_{y,D}^K(f)]$$

The coherency is commonly reported as "magnitude-squared coherence (MSC)" $|C(f)|^2$; the basic hypothesis test considers the null model for zero coherency (i.e., no correlation) between the two time series x and y versus the alternative model for nonzero coherency. Thus, when $|C(f)|^2$ is close to 1, the two series are correlated; specific frequencies may show strong correlation and others may show weak to no correlation (noise) with $|C(f)|^2 \to 0$. As discussed below, the properties of $|C(f)|^2$ indicate that there is strong uncertainty about what constitutes zero coherence and what does not and related to the dof of the estimation.

Basic application of these estimators is demonstrated in Figure 4.26 on two insolation models calculated for a ~0.8 Myr interval from 36.2 Ma to 37.0 Ma, one lagging the other by 5 kyr. The differences in noise content of the two models are quite evident in the power spectra: in Series 1, there is a hint of orbital eccentricity variation at frequencies 1/405, 1/128, and 1/95 kyr, which is not at all evident in the Series 2 spectrum. The effective dof are lower for the uniform noise background of Series 1. Otherwise, both models have high spectral peaks and maximum dof at the obliquity (1/41 kyr) and precession index (1/24, 1/22, 1/19, and 1/17 kyr) frequencies, which are strongly correlated, as borne out by the coherency estimates near 1.0. The cross-phase $\phi(f)$ shows a flattening of values across the frequency bands with high $|C(f)|^2$ that are consistent with a constant lag of 5-kyr of Series 2 with respect to Series 1. For example, at f = 22 kyr, the phase is +90°, which is one-fourth of a cycle, i.e., 22 kyr/4 = 5 kyr.

$|C(f)|^2$ has a probability distribution that was originally described for the general cross-correlation coefficient by R. A. Fisher and is given in Table 1 of Carter et al. (1973). The estimated value of $|C(f)|^2$ has a strong bias and variance as a function of dof K, shown in Figure 4.27, which must be accounted for in coherency estimation. Consequently, the level of "zero coherence" is not zero, and can be drawn at specific significance levels (Figure 4.27a), progressively lower for higher K (representative list in Table 4.1). Finally, these uncertainties propagate into the cross-phase

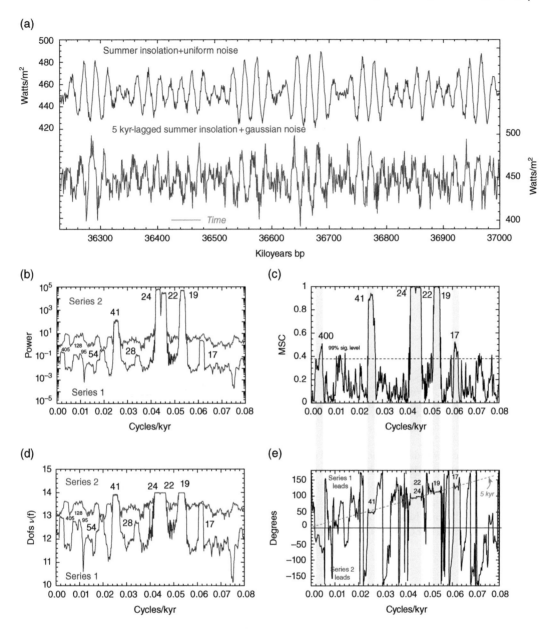

Figure 4.26 4π multitaper coherency and cross-phase spectral analysis of two insolation models for the interval 36.230–37.0 Ma (Arguis Formation time). (a) Two time series, sampled at Δt = 1 kyr, of 65°North summer half-year insolation: one with uniform noise and the other with Gaussian white noise, with Series 2 lagging Series 1 by 5 kyr (time proceeds from right to left). (b). Power spectra of the series with strong spectral peaks at the obliquity and precession index frequencies; Series 1, with its lower noise level, picks up the eccentricity frequencies, which contributes only 100th of the solar radiation variation in the Milankovitch cycles. (c). Effective dof for the spectral estimates of the two time series. (d). MSC with Series 1 entered first and Series 2 entered second into $|C(f)|^2$. (e). Cross-phase analysis; Series 1 leads in frequency with positive cross phase, and Series 2 leads in frequency with negative cross phase. The dashed green line indicates cross phase for a 5 kyr lead of Series 1 over Series 2. Calculated using *pmtm,m*, *mtmdofs.m*, and *mtmcoherency.m* (see Appendix).

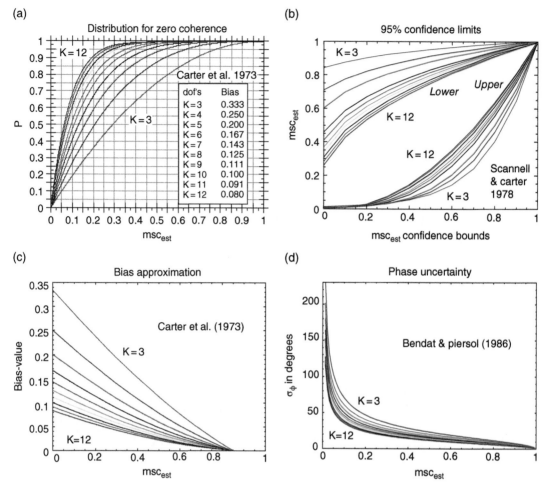

Figure 4.27 Statistics for squared-coherency and cross-phase spectra, computed using *mtmcoherencystats.m* (see Appendix). As a function of dof K (a) distribution for zero coherence where msc$_{est}$ on the horizontal axis can be related to the probability distribution P on the vertical axis; the table insert indicates the bias for zero coherency (see also c. at msc$_{est}$ = 0.0); (b) 95% CLs, where msc$_{est}$ on the vertical axis is related to lower and upper bounds as a function of K; (c) coherency bias to be subtracted from msc$_{est}$; (d) cross-phase uncertainty σ$_\phi$. Source: Data from Carter et al. 1973, Scannell & Carter 1978 and Bendat and Piersol 2010.

Table 4.1 CLs and significance levels for zero-squared coherence for K degrees of freedom

	95% confidence limits			Significance levels			
K	Lower	Upper	K	90%	95%	98%	99%
3	0.013000	0.84200	3	0.68400	0.77800	0.86100	0.90400
4	0.0080000	0.70800	4	0.53800	0.63500	0.73500	0.79500
5	0.0060000	0.60200	5	0.44000	0.53100	0.63300	0.70000
6	0.0040000	0.45900	6	0.37200	0.45600	0.55400	0.62300
7	0.0040000	0.41000	7	0.33200	0.39900	0.49200	0.56200
8	0.0030000	0.36900	8	0.28300	0.35400	0.44300	0.51200
9	0.0030000	0.33600	9	0.25300	0.31900	0.40300	0.47100
10	0.0030000	0.30800	10	0.22900	0.29000	0.37000	0.43800
11	0.0020000	0.28500	11	0.20900	0.26600	0.34200	0.41000
12	0.0020000	0.26500	12	0.19200	0.24600	0.31800	0.38700

(a)

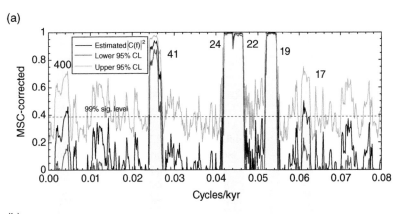

(b)

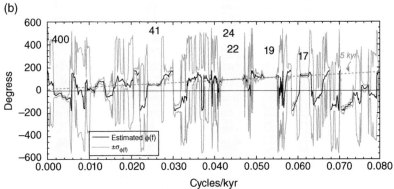

Figure 4.28 Statistics from Figure 4.27 and Table 4.1 applied to the coherency (a) and cross phase (b) spectral analysis shown in Figure 4.26. MSC peaks exceeding the lower 99% CL are considered to be significantly different from zero coherence. These statistics indicate that the MSC peaks at 1/400 kyr and 1/17 kyr does not pass this test due to interference from the noise.

estimates and are reported as standard deviation $\sigma\phi$ (Bendat & Piersol 1986). Application of these statistics to the example in Figure 4.26 is shown in Figure 4.28 and reveals that some of the coherency peaks are not significantly different from zero.

References

Abe, M. & Smith, III, J.O. (2004) Design criteria for the quadratically interpolated FFT method (I): Bias due to interpolation, in *Technical Report STAN-M-114*, p. 13. Department of Music, Stanford University, Stanford.

Bartlett, M.S. (1948) Smoothing periodograms from time series with continuous spectra. *Nature*, *191*, 686–687. DOI:10.1038/161686a0.

Bartlett, M.S. (1950) Periodogram anlaysis and continuous spectra. *Biometrika*, *37*, 1–16. DOI:10.1093/biomet/37.1-2.1.

Bendat, J.S. & Piersol, A.G. (1986) *Random Data: Analysis and Measurement Procedures. Second Edition*. Wiley and Sons, New York.

Blackman, R.B. & Tukey, J.W. (1958) *The Measurement of Power psectral from the Point of View of Communication Engineering*. Dover, New York.

Carter, G.C., Knapp, C.H., & Nuttall, A.H. (1973) Estimation of the magnitude-squared coherence function via overlapped Fast Fourier Transform processing. *IEEE Transactions, Audio and Electroacoustics, AU-21*, 337–344. DOI:10.1109/TAU.1973.1162496.

Cooley, J.W. & Tukey, J.W. (1965) An algorithm for the machine calculation of complex Fourier series. *Mathematics of Computation, 19*, 297–301. DOI:10.1090/S0025-5718-1965-0178586-1.

Durrani, T.S. & Nightingale, J.M. (1972) Data windows for digital spectral analysis. *Proceedings of the IEEE, 119*, 343–352. DOI:10.1049/piee.1972.0080.

Elgar, S. (1987) Bias of effective degrees of freedom. *Journal of Waterway, Port, Coastal and Ocean Engineering, 113*, 77–82. DOI:10.1061/(ASCE)0733-950X (1987)113:1(77).

Ghil, M., Allen, M.D., Dettinger, K.I., Kondrashov, D., Mann, M., Roberts, A.P., Saunders, A., Tian, Y., Varadi, F., & Yiou, P. (2002) Advanced spectral methods for climatic time series. *Rev. Geophys., 40*, 3.1–3.41. DOI:10.1029/2000RG000092. DOI:10.1029/2000RG000092.

Gillispie, D.T. (1996) The mathematics of Brownian motion and Johnson noise. *American Journal of Physics, 64*, 225–240. DOI:10.1119/1.18210.

Gilman, D.L., Fuglister, F.J., & Mitchell, J.M. (1963) On the power spectrum of "red noise". *Journal of Atmospheric Sciences, 20*, 182–184. DOI:10.1175/1520-0469 (1963)020<0182:OTPSON>2.0.CO;2.

Harris, F.J. (1978) On the use of window for harmonic analysis with the discrete Fourier transform. *Proceedings of the IEEE, 66*, 51–83. DOI:10.1109/PROC.1978.10837.

Hinnov, L.A., Olson, P.L., & Driscoll, P.E. (2012) Sherman statistic reveals nonrandom behavior in the Phanerozoic Polarity Time Sclae, *AGU Fall Meeting*, San Francisco.

Hinnov, L.A., Schulz, M., & Yiou, P. (2002) Interhemispheric space-time attributes of the Dansgaard-Oescher oscillations between 0-100 ka. *Quaternary Science Reviews, Special Volume: Decadal to Millennial Climate Change, 21*, 1213–1228. DOI: 10.1016/S0277-3791(01)00410-8

Jenkins, G.M. & Watts, D.G. (1968) *Spectral Analysis and Its Applications*, 525 pp. Holden-Day, San Francisco.

Klapper, J. & Harris, C.M. (1959) On the response and approximations of Gaussian filters. *Institute of Radio Engineers Transactions on Audio, AU-4* (3), 80–87. DOI:10.1109/TAU.1959.1166198

Kodama, K.P., Anastasio, D.J., Newton, M.L., Pares, J., & Hinnov, L.A. (2010). High-resolution rock magnetic cyclostratigraphy in an Eocene flysch, Spanish Pyrenees. *Geochemistry, Geophysics, Geosystems, 11*. DOI:10.1029/2010GC003069.

Mann, M. & Lees, J. (1996) Robust estimation of background noise and signal detection in climatic time series. *Climate Change, 33*, 409–445. DOI:10.1007/BF00142586.

Meyers, S.R. (2012) Seeing red in cyclic stratigraphy: spectral noise estimation for astrochronology. *Paleoceanography, 27* (*PA3228*). DOI:10.1029/2012PA002307.

Milotti, E. (2002) *1/f Noise: A Pedagogical Review*, http://arxiv.org/abs/physics/0204033, accessed on May 28, 2014.

Olson, P.L., Hinnov, L.A., & Driscoll, P.E. (2013) Nonrandom geomagnetic polarity reversal times and geodynamo evolution. *Earth and Planetary Science Letters, 388*, 9–17. DOI:10.1016/j.epsl.2013.11.038.

Paillard, D., Labeyrie, L., & Yiou, P. (1996) Macintosh program performs time-series analysis. *Eos, Transactions American Geophysical Union, 77*, 379. DOI:10.1029/96EO00259.

Park, J., Lindberg, C.R., & Vernon III, F.L. (1987) Multitaper spectral analysis of high-frequency seismograms. *Journal of Geophysical Research, 92*, 12675–12684. DOI:10.1029/JB092iB12p12675.

Percival, D.B. & Walden, A.T. (1993) *Spectral Analysis For Physical Applications*. Cambridge University Press, Cambridge.

Press, W.H. (1978) Flicker noises in astronomy and elsewhere. *Comments on Astrophysics, 7*, 103–119. A&AA ID. AAA022.061.024

Priestley, M.B. (1981) *Spectral Analysis and Time Series*, 925 pp. Academic Press, San Diego.

Scannell, E.H. & Carter, G.C. (1978) Confidence bounds for magnitude-squared coherence estimates. *IEEE Transactions on Acoustics, Speech and Signal Processing, ASSP-26*, 475–477. DOI:10.1109/TASSP.1978.1163131.

Schulz, M. & Mudelsee, M. (2002) REDFIT: Estimating red-noise spectra directly from unevenly spaced palaeoclimatic time series. *Computers and Geosciences, 28*, 412–426. DOI:10.1016/S0098-3004(01)00044-9.

Schulz, M. & Schaltegger, K. (1997) SPECTRUM: Spectral analysis of univernly sampled palaeoclimatic time series. *Computers and Geosciences, 23*, 929–945. DOI:10.1016/S0098-3004(97)00087-3.

Schuster, A. (1897) On lunar and solar periodicities of earthquakes. *Proceedings of the Royal Society of London, 61*, 455–465. DOI:10.1098/rspl.1897.0060.

Slepian, S. (1978) Prolate spheroidal wave functions, Fourier analysis and uncertainty-V: The discrete case. *Bell Systems Technical Journal, 57*, 1371–1430. DOI:10.1002/j.1538-7305.1978.tb02104.x.

Taner, M.T. (2000) *Attributes Revisited*, Technical Report. Rock Solid Images, Inc. http://www.rocksolidimages.com/pdf/attrib_revisited.htm, accessed May, 2013.

Taner, M.T., Koehler, F., & Sheriff, R.E. (1979) Complex trace analysis. *Geophysics, 44*, 1041–1063. DOI:10.1190/1.1440994.

Thomson, D.J. (1977) Spectrum estimation techniques for characterization and development of WT4 waveguide-II. *Bell Systems Technical Journal, 56*, 1983–2005. DOI:10.1002/j.1538-7305.1977.tb00164.x.

Thomson, D.J. (1982) Spectrum estimation and harmonic analysis. *IEEE Proc, 70*, 1055–1096. DOI:10.1109/PROC.1982.12433.

Thomson, D.J. (1990) Quadratic-inverse spectrum estimates: applications to palaeoclimatology. *Philosophical Transactions of Royal Society A, 332*, 539–597. DOI:10.1098/rsta.1990.0130.

Thomson, D.J. (2009) Time-series analysis of paleoclimate data. In: Gornitz, V. (ed), *Encyclopedia of Paleoclimatology and Ancient Environments*, pp. 949–959. Springer, New York.

Voss, R.F. (1979) 1/f (flicker) noise: A brief review. In: *IEEE 33rd Annual Symposium on Frequency Control*, pp. 40–46. DOI:10.1109/FREQ.1979.200297.

Welch, P.D. (1967) The use of FAst Fourier transform for the estimation of power spectra: A method based on time averagin over short, modified periodograms. *IEEE Transactions, Auido and Electroacoustics, AU-15*, 70–73. DOI:10.1109/TAU.1967.1161901.

5 Milankovitch Forcing Theory

Abstract: This chapter reviews the basic theory of astronomical forcing of paleoclimate that was originally described by Milutin Milankovitch, with a summary of the frequencies of Earth's orbital eccentricity, obliquity (tilt), and precession index. The astronomically forced insolation is explained and analyzed with a focus on the seasonal aspects of the forcing. The chapter concludes with a description of practices for developing cyclostratigraphic timescales, with the traditional tuning of an Eocene section.

5.1 Introduction

Once time series analysis (Chapter 4) has identified the dominant frequencies in a cyclostratigraphic sequence, the real challenge is determining if any are Milankovitch cycles in scale and in origin. This chapter reviews the Milankovitch forcing theory of climate (Milankovitch 1941) and how Milankovitch cycles are identified through time series analysis. Tuning techniques and approaches are also covered here. Developing a refined timescale is accomplished through application of tuning, identifying likely Milankovitch frequencies, and adjusting errors in the frequencies that are not attenuated by the initial timescale.

5.2 Astronomical Parameters

Variations in the Earth's orbit and tilt relative to the Sun force changes in seasonal, latitudinal, and total insolation, leading to global climate cycles at 10^4–10^6 year timescales. There are two categories of variations, those arising from solar system dynamics (the orbital elements) and those arising from Earth–Moon dynamics (Earth rotation and shape, obliquity of the ecliptic, and precession rate) (Figure 5.1).

Rock Magnetic Cyclostratigraphy, First Edition. Kenneth P. Kodama and Linda A. Hinnov.
© 2015 John Wiley & Sons, Ltd. Published 2015 by John Wiley & Sons, Ltd.

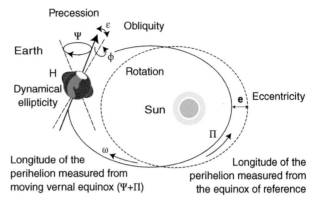

Figure 5.1 Astronomical parameters affecting Earth–Sun position. Source: Hinnov and Ogg 2007.

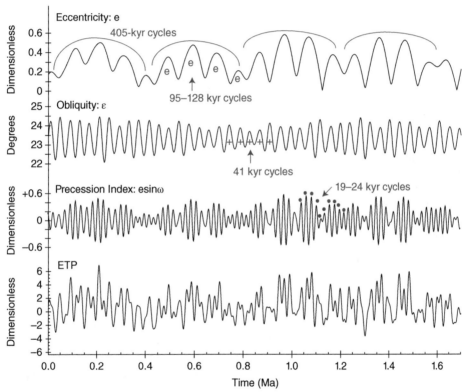

Figure 5.2 The astronomical parameters 10^4–10^5 years oscillations for the past 1.6 million years according to the La2004 nominal solution (Laskar et al. 2004). Main periodicities in kiloyears are indicated in red. ETP is the sum of the standardized eccentricity E, obliquity (i.e., tilt T) and precession index P (Imbrie et al. 1984). Source: Hinnov and Ogg 2007.

The parameters that affect climate are (i) Earth's orbital eccentricity e, which determines Earth–Sun distance, (ii) Earth's axial tilt, or obliquity ε, and (iii) the precession of Earth's rotation axis or the **precession of the equinoxes** ψ, which determines the timing and location of the seasons with respect to the Earth's orbit. Numerical solutions (e.g., Laskar et al. 2004, 2011) have been calculated for hundreds of millions of years into the past, revealing complex modulating behavior in the astronomical parameters (Figures 5.2 and 5.3). The orbital

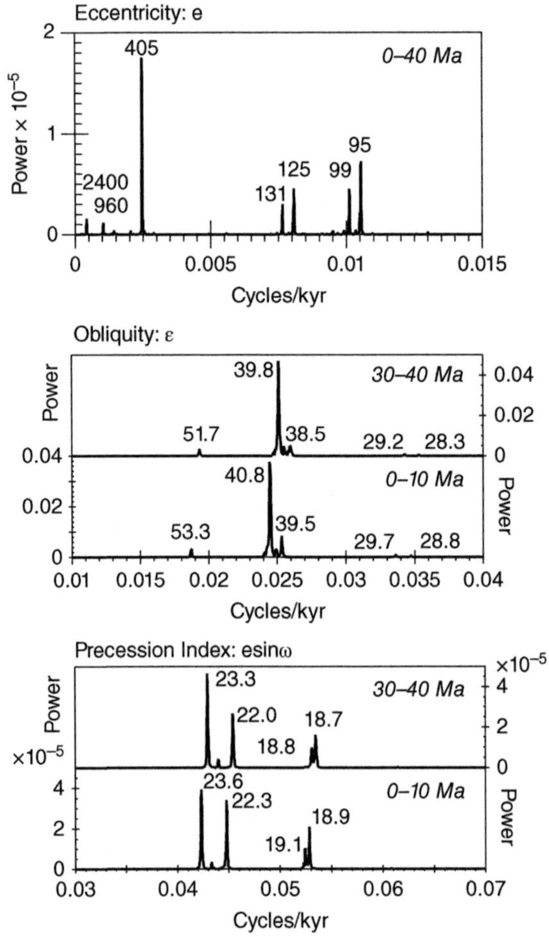

Figure 5.3 –2π prolate multitapered power spectra of the La2004 astronomical parameters over the past 40 million years. Spectra for 0–10 Ma and 30–40 Ma are calculated for the obliquity and precession index to demonstrate the shift to higher frequencies (shorter periodicities) due to the tidal dissipation effect. Source: Hinnov 2013. Reproduced with permission of Geological Society of America.

eccentricity has varied between ~0 and 0.06, with major periodicities in the ~100 kyr band and at 405 kyr. The *obliquity* has ranged between 22.5° and 24.5°. Faster rotation rates from tidal dissipation shifts variations to shorter periods in earlier time. The *precession index* tracks the drift of the seasons and relative Earth–Sun distance around the Earth's orbit relative to perihelion due to the Earth's precession. This drift, along with the **precession of the orbital perihelion** (Π) in the opposite direction plus changes in orbital eccentricity, produces the modulated precession index (esinω).

The terms "orbital forcing" and "astronomical forcing" have become synonymous in the literature. In fact, "orbital" is not technically correct, in that the obliquity and precession index do not have strict orbital origins but also involve geophysical phenomena (Hinnov 2013). The terms "precessional parameter," "**climatic precession**," "astroclimatic parameter," and "**precession-eccentricity syndrome**" are all synonyms for the precession index. ETP (Figure 5.2) is often adopted in cyclostratigraphic–astronomical forcing studies where details of the forcing, e.g., the magnitude and/or phase response of the forcing, are not understood.

5.3   Insolation

The astronomical parameters affect changes in the intensity and timing of the incoming solar radiation, or insolation, at all points on the Earth. The insolation variations are the well-known Milankovitch cycles. Geographical location, time of year, and time of day together determine the relative contributions of the orbital parameters to the interannual insolation (e.g., Berger et al. 1993, 2010). Climate integrates insolation over certain times of the year and collectively over specific geographic areas, possibly over different areas at different times. This "climatic filtering" alters the relative contributions of the orbital parameters to the total output climate response, this even prior to internal climate system responses to the insolation. Thus, it is left to the discretion of the paleoclimatologist to determine which time(s) of the year and at which location(s) a prevailing climate has responded to insolation.

A still largely unrecognized aspect of insolation is the interannual phasing of the astronomical parameters as a function of time of year. For example, at 35° North (Figure 5.4), the obliquity is in phase with insolation during the summer months, but antiphased during the winter months. Precession-forced insolation continuously changes phase with respect to the standard precession index over the course of a year. Eccentricity forcing, while extremely low in power, is in phase with insolation throughout the year, i.e., higher eccentricity corresponds to higher insolation.

5.4   Astronomical Tuning and Timescales

Tuning stratigraphic sequences can have different meanings, and it is always controversial in one way or another, particularly "astronomical tuning." In this section, a detailed example of traditional astronomical tuning is presented, with discussion of the limitations imposed by the tuning in terms of interpretation. This is followed by a discussion of a new class of tuning methods that are coming online, based on searches for the sedimentation rate that statistically produces the best fit to an astronomical model. Concepts and procedures are demonstrated on the Eocene Arguis Formation (Kodama et al. 2010) (Figure 5.5).

5.4.1   The Initial Timescale

Assigning an initial timescale is a required step in determining whether Milankovitch cycles are present in cyclostratigraphy. The usual tools involve some combination of magnetostratigraphy, biostratigraphy, chemostratigraphy, and radioisotope dating, depending on what is available. In the case of the Arguis Formation, biostratigraphy indicates that the formation spans nannoplankton (NP) zones NP16–NP17 and planktonic foraminifera zones

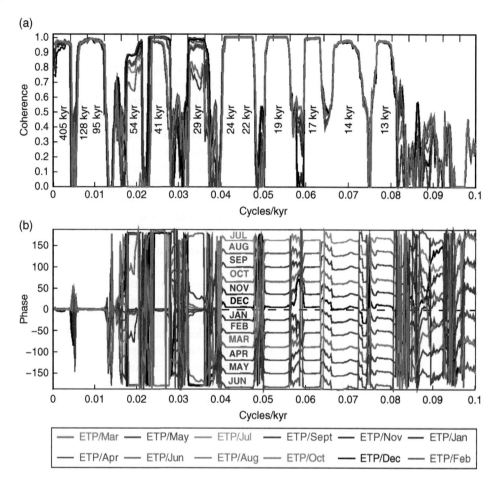

Figure 5.4 4π prolate multitapered coherency (a) and cross-phase (b) spectra of the ETP versus monthly insolation at 35° North from 40 to 36 Ma, time increasing toward the present. ETP and monthly insolation were computed using *Analyseries 2.0.4.2* (Paillard et al. 1996); spectra were computed with *cmtm.m* by Peter Huybers (see Appendix). High coherency and stable cross phase is detected in the eccentricity, obliquity, and precession index bands, as well as in minor, high-frequency precession index bands at $(14\ \text{kyr})^{-1}$ and $(13\ \text{kyr})^{-1}$. Source: Kodama, Anastasio, Newton, Pares & L. A. Hinnov 2010. Reproduced with permission of John Wiley & Sons, Inc.

Figure 5.5 Timescales for the Arguis ARM cyclostratigraphy. 2π prolate multitaper power spectra on the left refer to untuned and tuned ARM series on the right. (a) Original ARM series in the stratigraphic domain. (b) The ARM time series according to GTS2004 magnetic reversal ages (GPTS2004). (c) The 405 kyr–tuned ARM time series, with 405 kyr tie points indicated by vertical blue lines. The lowpass Taner filter isolates 405 kyr cyclicity. (d) La2004 eccentricity–tuned time series, first iteration, shown with blue vertical lines between ARM minima and eccentricity minima. (e) La2004 eccentricity tuning, second iteration. Minima and maxima of the lowpass filtered ARM series are tied to minima and maxima of the eccentricity series, shown with blue vertical tie lines. (f) La2004 precession index tuning of bandpassed precession band of the eccentricity, second iteration tuned ARM series. (g) Depth-time transformation from each tuning. (h) ARM series referring to the power spectra; the bottom-most series is the La2004 precession index–tuned accumulation rate series defined at ~10 kyr intervals. The time picks were made using the "linage" function in Analyseries (Paillard et al. 1996); see also *picktune.m* (Appendix).

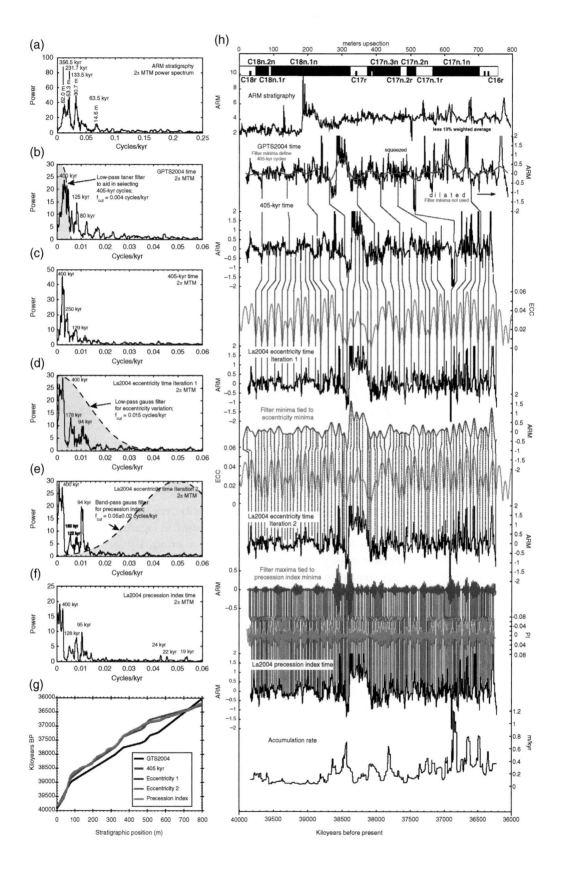

P12–P15, constraining the magnetic reversal stratigraphy of the formation to Chrons C16r–C18r. This provides an initial timescale to be inferred using the GPTS (Geomagnetic Polarity Time Scale) provided in the Geologic Time Scale 2004 (Gradstein et al. 2004), from 39,898.816 to 36,065.988 ka. The anhysteretic remanent magnetization (ARM) series tuned to GPTS2004 is shown in Figure 5.5h; its power spectrum in Figure 5.5b.

5.4.2   Traditional Astronomical Tuning

Traditional astronomical tuning is a "user-intensive" procedure that involves directly matching cyclostratigraphic variations to an astronomical model. This is where much of the criticism arises, because the procedure can (i) erroneously confine variability into the astronomical frequencies and (ii) lead to circular reasoning. Therefore, careful calculations must be made to avoid or minimize these problems and to judge what the final objectives are for the tuning.

For example, the Plio–Pleistocene sapropels of the Mediterranean were tuned to a full astronomical model incorporated in a 65° North summer insolation curve (Lourens et al. 1996, 2004; Hilgen et al. 2012). The objective was to define an **astrochronology** for the Plio–Pleistocene epochs based on the success of fit between the sapropels and insolation maxima, guided by the magnetostratigraphic timescale of Cande and Kent (1992, 1995). This procedure can be used safely to interpret the goodness of fit between astronomical target and sapropel pattern; independently determined geomagnetic polarity reversal ages can support the outcome.

One solution is to tune only one frequency, or one parameter, a procedure known as "minimal tuning" (Muller & MacDonald 2000). The results of minimal tuning, to the 405-kyr eccentricity cycle, and then to the full eccentricity solution (in two iterations), are shown in Figure 5.5c–e and h. Already in the GPTS2004-tuned ARM series, a strong 400 kyr scale cyclicity is evident through the series, and from this, a new 405 kyr–tuned ARM series was constructed. The tuning could be stopped at this point (and for stratigraphy older than 50 Ma, it must; Laskar et al. 2011). However, uncorrected sedimentation rate fluctuations likely remain that distort the timescale within the 405 kyr tie points.

A more radical minimal tuning may be considered that matches the 405 kyr–tuned ARM series to the full orbital eccentricity solution. For the ARM series, two iterations were undertaken: the first tuned ARM minima to eccentricity minima; the second iteration tuned a low-pass version of the ARM first iteration eccentricity time series to the maxima and minima of the same orbital eccentricity solution. This procedure sharpens the peak at 100 kyr and progressively suppresses the non-orbital peak at 250 kyr, to a weaker peak at 178 kyr, and finally to a minor peak at 180 kyr. At this point, the eccentricity-tuned ARM time series can still be somewhat safely examined for the presence of obliquity and precession forcing (Kodama et al. 2010; see also Chapter 6).

The final tuning presented here, namely tuning the ARM-filtered precession signal to the precession index in the mid-summer phase (i.e., ARM maxima to precession index minima), was undertaken in an attempt to improve the eccentricity spectrum, which appears to be accomplished (Figure 5.5f). The resulting sedimentation rates (Figure 5.5h, bottom series) show strong 405-kyr variations, which amplify and include ~100-kyr cyclicity toward the top of the series.

5.4.3 Objective Astronomical Tuning

Objective techniques have recently been developed to model the timescale problem using statistics. The "average spectral misfit" method (Meyers & Sageman 2007; Meyers 2008) identifies the sedimentation rate that best transforms a stratigraphic spectrum to a model astronomical spectrum by comprehensively testing a range of likely sedimentation rates on a stratigraphic series and assessing the output spectra with respect to a model astronomical spectrum. The sedimentation rate with the lowest number of fits to Monte Carlo–generated randomized spectra is taken as the most likely solution. A similar "Bayesian Monte Carlo" approach was developed by Malinverno et al. (2010), which searches for the sedimentation rate that maximizes a likelihood function defined by the ratio of data and equivalent red noise spectrum weighted by astronomical frequencies. Both methods provide statistics on the tested sedimentation rates and preserve the original phasing of the data, a requirement for interpretation of insolation forcing and for evaluating lags of astronomical-forced proxies with respect to a model.

References

Berger, A., Loutre, M.F., & Tricot, C. (1993) Insolation and Earth's orbital periods. *Journal of Geophysical Research*, *98*, 10341–10362. DOI:10.1029/93jd00222.

Berger, A., Loutre, M.F., & Yin, Q.Z. (2010) Total irradiation during any time interval of the year using elliptic integrals. *Quaternary Science Reviews*, *29*, 1968–1982. DOI:10.1016/j.quascirev.2010.05.007.

Cande, S.C. & Kent, D.V. (1992) A new geomagnetic polarity time scale for the Late Cretaceous and Cenozoic. *Journal of Geophysical Research*, *97*, 13917–13951. DOI:10.1029/92jb01202.

Cande, S.C. & Kent, D.V. (1995) Revised calibration of the geomagnetic polarity timescale of the Late Cretaceous and Cenozoic. *Journal of Geophysical Research*, *100*, 6093–6095. DOI:10.1029/94jb03098.

Gradstein, F.M., Ogg, J.G., & Smith, A. (2004) *A Geologic Time Scale*. Cambridge University Press, New York.

Hilgen, F., Lourens, L.J., & Van Dam, J.A. (2012) Chapter 29: The Neogene period. In: Gradstein, F.M., Ogg, J.G., Schmitz, M.D., & Ogg, G.M. (eds), *The Geologic Time Scale 2012*, pp. 921–978. Elsevier, Amsterdam.

Hinnov, L.A. (2013) Cyclostratigraphy and its revolutionizing applications in the Earth and planetary sciences. *Geological Society of America Bulletin*, 125,11-12;1703-1734. DOI: 10.1130/B30934.1.

Hinnov, L.A. & Ogg, J.G. (2007) Cyclostratigraphy and the astronomical time scale. *Stratigraphy*, *4*, 239–251.

Imbrie, J., Hays, J.D., Martinson, D.G., McIntyre, A., Mix, A.C., Morely, J.J., Pisias, N.G., Prell, W.L., & Shackleton, N.J. (1984) The orbital theory of Pleistocene climate: Support from a revised chronology of the marine d18O record. In: Berger, A.L., Imbrie, J., Hays, J.D., Kukla, G., & Saltzman, B. (eds), *Milankovitch and Climate, Part 1*, pp. 269–305, Reidel Publishing, Dordrecht.

Kodama, K.P., Anastasio, D.J., Newton, M.L., Pares, J., & Hinnov, L.A. (2010) High-resolution rock magnetic cyclostratigraphy in an Eocene flysch, Spanish Pyrenees. *Geochemistry, Geophysics, Geosystems*, *11*. DOI:10.1029/2010GC003069.

Laskar, J., Robutel, P., Joutel, F., Gastineau, M., Correia, A.C.M., & Levrard, B. (2004) A long term numerical solution for the insolation quantitties of the Earth. *Astronomy and Astrophysics*, *428*, 261–285. DOI:10.1051/0004-6361:20041335.

Laskar, J., Fienga, A., Gastineau, M., & Manche, H. (2011) La2010: A new orbital solution for the long-term motion of the Earth. *Astronomy and Astrophysics*, *532* (*A89*). DOI:10.1051/0004-6361/201116836.

Lourens, L.J., Antonarakou, A., Hilgen, F., Van Hoof, A.A.M., & Vergnaud-Grazzini, C. (1996) Evaluation of the Plio-Pleistocene astronomical timescale. *Paleoceanography*, *11* (4), 391–413. DOI:10.1029/96PA01125.

Lourens, L.J., Hilgen, F., Shackleton, N.J., Laskar, J., & Wilson, D. (2004) The Neogene period. In: Gradstein, F.M., Ogg, J.G., & Smith, A. (eds), *A Geologic Time Scale 2004*, pp. 400–440. Cambridge University Press, Cambridge.

Malinverno, A., Erba, E., & Herbert, T.D. (2010) Orbital tuning as an inverse problem: Chronology of the early Aptian oceanic anoxic event 1a (Selli Level) in the Cismon APTOCORE. *Paleoceanography*, *25*(*PA2203*). DOI:10.1029/2009PA001769.

Meyers, S.R. (2008) Resolving Milankovitchian controversies: The Triassic Latemar Limestone and the Eocene Green River Formation. *Geology*, *36*, 319–322. DOI:10.1130/G24423A.1.

Meyers, S.R. & Sageman, B.B. (2007) Quantification of deep-time orbital forcing by average spectral misfit. *American Journal of Science*, *307*, 773–792. DOI:10.2475/05.2007.01.

Milankovitch, M. (1941) *Kanon der Erdbestrahlung und seine Anwendung auf das Eiszeitproblem*, 633 pp. Royal Serbian Academy, Secton of Mathematical and Natural Sciences, Belgrade.

Muller, R.A. & MacDonald, G.J. (2000) *Ice Ages and Astronomical Causes: Data, Spectral Analysis, and Mechanisms*. Springer-Praxis, London.

Paillard, D., Labeyrie, L., & Yiou, P. (1996) Macintosh program performs time-series analysis. *Eos, Transactions American Geophysical Union*, *77*, 379. DOI:10.1029/96eo00259.

6

Case Studies of Rock Magnetic Cyclostratigraphy

Abstract: Details of rock magnetic cyclostratigraphy studies are given in this chapter to illustrate how the technique is applied to sedimentary sequences deposited in a variety of environments, including shallow marine, platform carbonate, and fluvial settings. These case studies show that rock magnetic cyclostratigraphy can detect astronomically forced climate cycles in rocks that vary in age from Plio-Pleistocene to Neoproterozoic. The rock magnetics appear to encode climate variations by different techniques, including eolian dust driven by aridity or continental flux driven by runoff.

6.1 Introduction and Environmental Shredding

This chapter will present examples of rock magnetic cyclostratigraphy studies to illustrate the details of conducting a study and show how those details can vary on a case-by-case basis. The general guidelines for conducting a rock magnetic cyclostratigraphy study listed in Chapter 1 are, just that, guidelines. As in any creative enterprise, the researcher must adjust their techniques and methodologies to fit the specific characteristics of the rocks being studied in order to answer the important question: are the cycles observed truly due to astronomically forced climate cycles? As the examples in this chapter will show, the rock magnetic parameter that gives the best result, whether its susceptibility, anhysteretic remanent magnetization (ARM), or isothermal remanent magnetization (IRM) or some ratio of parameters, will need to be tailored to the rocks being studied. The magnetic minerals carrying the signal also need to be identified and their age of creation needs to be pinned down, so the mechanism by which the astronomically forced cycles are encoded can be understood. Unraveling the encoding mechanism is an important goal for any rock magnetic cyclostratigraphy study. Not only does it lend credence to the identification

Rock Magnetic Cyclostratigraphy, First Edition. Kenneth P. Kodama and Linda A. Hinnov.
© 2015 John Wiley & Sons, Ltd. Published 2015 by John Wiley & Sons, Ltd.

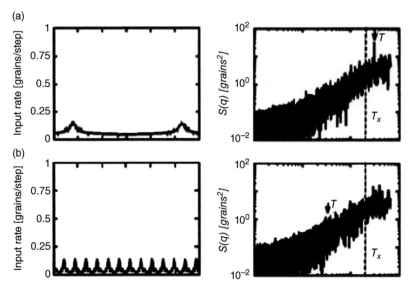

Figure 6.1 Figure from Jerolmack and Paola (2010) showing their modeling of numerical rice-pile experiments in which the input signal (with period T, (a)) has a longer period than the characteristic frequency of the rice-pile processes (landsliding primarily, Tx) or a shorter period (b). In the bottom example, the amplitude of the input signal keeps it from being completely shredded. Source: Jerolmack & Paola 2010. Reproduced with permission of John Wiley & Sons, Inc.

of the cycles, but it teaches us the important details about surficial processes in the Earth system and is a satisfying intellectual reward.

Another important consequence of examining specific examples of rock magnetic cyclostratigraphy studies is to illustrate the limitations of the technique and the important role of "shredding" of the input signal by the sediment transport processes for a given depositional environment. Jerolmack and Paola (2010) propose that the sediment transported through a landscape will be affected by processes that act, in concert, as a nonlinear filter that can "shred" an environmental signal. Jerolmack and Paola (2010) point out that processes such as landsliding, bed load transport, and river avulsion will act as a **low-pass filter** for the environmental signal input into the system. Their modeling of rice-pile analogues of sedimentary systems suggests that if the frequency of the input signal is longer than the periods of the shredding processes, then the signal will make it through the system and be recorded in the stratigraphy (Figure 6.1). However, for input signals with periods shorter than a "threshold frequency," the input signal will be lost, unless the amplitude of the input signal is very large. Furthermore, the size of the depositional system, e.g., the size of the fluvial system carrying the sediment to the depositional basin, determines the "threshold frequency" that characterizes the system.

Jerolmack and Paola's (2010) idea of environmental "shredding" must be considered when interpreting the origin of the cycles observed in any rock magnetic cyclostratigraphic study. Furthermore, the successful identification of eccentricity and precession-scale signals in a rock magnetic record helps constrain the size of the threshold frequency for a given depositional environment. Jerolmack and Paola's (2010) model should also inform the researcher in the choice of depositional systems to investigate and estimates for the chances of success. For instance, the expected nearly continuous sedimentation of a near-shore marine depositional environment may have a better chance of recording an astronomically forced signal than the discontinuous sedimentation expected for a continental, fluvial depositional environment. However, if Milankovitch-scale cycles are identified in an ancient fluvial environment, the timescale of potential shredding processes is constrained and more is learned about Earth processes.

In this chapter, the case studies will be presented back through geologic time with the youngest studies first. We will cover rock magnetic cyclostratigraphic studies of Plio-Pleistocene marine sediments at the Stirone River section, northern Italy, Eocene marine marls of the Arguis Formation, Spanish Pyrenees, Cretaceous platform carbonates of the Cupido Formation, northeastern Mexico, Triassic platform carbonates of the Latemar, in the Dolomites of northern Italy, Triassic marine limestones and shales of the Daye Formation, South China, Mississippian fluvial red beds of the Mauch Chunk Formation, northeastern Pennsylvania, and the siltstones and carbonates of the Neoproterozoic Rainstorm Member of the Johnnie Formation, Death Valley, California.

6.2   Stirone River Section, Northern Italy

The Stirone River section exposes over 600 m of Plio-Pleistocene marine sediments in the Po Valley in the Apennines of northern Italy. The section includes three formations (Colombacci Fm., Argille Azzurre Fm., and Stirone Fm.) in a coarsening upward sequence of growth strata. Lithologically, the Stirone River section is comprised of marine mudstones, siltstones, and sandstones. The rock magnetic cyclostratigraphy was completed by Gunderson et al. (2012) on the 342 m of the Stirone Formation which is composed primarily of fossiliferous silty mudstones. Mary et al. (1993) and Channell et al. (1994) previously studied the magnetostratigraphy and rock magnetism of the Stirone River section. In addition to finding a reversal stratigraphy of Plio-Pleistocene age, they found significant evidence of secondary iron sulfides as well as primary depositional magnetite. Fossils in the section also help pinpoint the rock's age.

The Gunderson et al. (2012) study had two parts: 74 oriented samples were collected at 21 horizons for a magnetostratigraphic study and unoriented samples were collected every meter from 311 m of section for a rock magnetic cyclostratigraphic study. The ultimate goal of the cyclostratigraphic study was to provide a high-resolution age model for the growth strata. The age model of the growth strata would time the folding of the Salsomaggiore anticline which, in turn, would give a high-resolution measure of fault slip for the blind thrust driving the anticline upward.

Alternating field demagnetization of the oriented samples revealed a N-R-N-R-N polarity stratigraphy that could be correlated to the youngest part of the Gauss (Chron C2A) and the oldest part of the Matuyama (Chron C2) with the top normal polarity interval correlated to the Jaramillo event (C1r.1n; Gunderson et al. 2012). The tie to the geomagnetic polarity time scale (GPTS) indicates that the section ranges from 3 to 1 Ma in age and has an average sediment accumulation rate of 24 cm/kyr that was used as a first approximation to determine the duration of the cycles observed by the rock magnetics.

Magnetic susceptibility (MS), and ARM were measured for the unoriented cyclostratigraphic samples. IRM acquisition results suggested two coercivity components identified by Gunderson et al. (2012) as low coercivity (~20 mT) depositional magnetite and higher coercivity (~80 mT) iron sulfides. The iron sulfide component was much stronger than the magnetite component, so iron sulfides control a good portion of the ferromagnetic signal and hence could potentially obscure the cyclostratigraphy.

Multitaper method (MTM) spectral analysis showed that the susceptibility record was best at recording cycles in the data. Spectral peaks were observed with periods of 85 m, 10 m, and two peaks near to 5–6 m (Figure 6.2). The sediment accumulation rate derived from the magnetostratigraphy suggested that

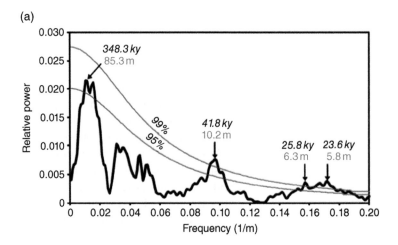

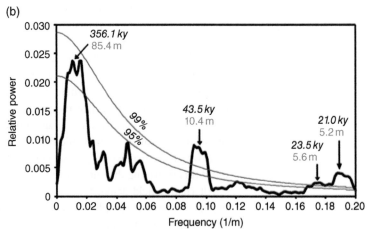

Figure 6.2 MTM power spectra for the Stirone Formation MS data series. (a) is untuned and (b) is tuned to theoretical obliquity. Source: Gunderson, Kodama, Anastasio & Pazzaglia 2012. Copyright 2012 by the Geological Society of London.

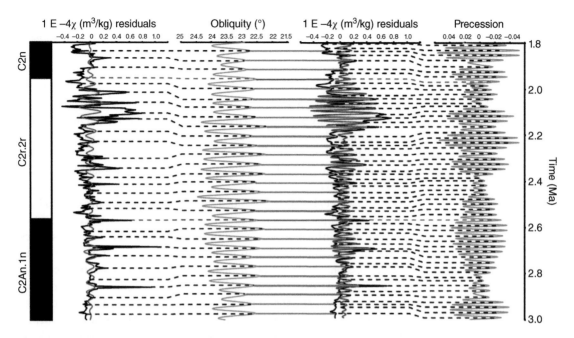

Figure 6.3 Figure taken from Gunderson et al. (2012) showing how detrended MS data (low frequencies removed, black curve on left) is tied to theoretical obliquity (blue curve second from left). The obliquity-tuned data series is then Gaussian filtered at precession scale (red curve second from right) and tied to theoretical precession (blue curve, far right). Source: Gunderson, Kodama, Anastasio & Pazzaglia 2012. Copyright 2012 by the Geological Society of London.

the 10.2 m peak (41.8 kyr) would be close to obliquity in duration. Therefore, the susceptibility record was tuned to theoretical obliquity for the Plio-Pleistocene and the 5–6 m peaks emerged as precession with periods of 23.5 and 21 kyr. The tuning also leads to a spectral peak near to 95 kyr which could be short eccentricity, but it does not reach the 95% confidence limits above the robust red noise calculated for the time series using the algorhythm of Mann and Lees (1996). A strong spectral peak does arise at about 350 kyr, but Gunderson et al. (2012) do not identify it as being astronomically forced.

One interesting story that is not in Gunderson et al.'s (2012) paper is that the data were collected in two field seasons, with some of the oriented samples collected in each season and all the unoriented samples collected in the first season. Based on the identification of obliquity in the MS data series, Kellen Gunderson predicted where the base of the Olduvai event (C2n) and the Gauss–Matuyama boundary (C2r.1n-C2An.1n) should be found in the stratigraphic section. Sampling during the second field season in this part of the section found the reversal boundaries exactly as predicted, a strong testament to the accurate identification of the obliquity cycle.

The final high-resolution age model was produced by band-pass filtering the obliquity-tuned data series at the precession scale (mean period = 21.3 kyr) and retuning the record to theoretical precession (Figure 6.3).

Understanding the encoding of the obliquity and precession scale orbitally forced signals in the Stirone Formation by susceptibility is problematic simply because susceptibility can be carried by ferromagnetic, paramagnetic, and diamagnetic minerals. Furthermore, many of the ferromagnetic minerals are secondary iron sulfides, as shown by Gunderson et al.'s (2012) work and earlier work by Channell et al. (1994) and Mary et al. (1993). Measurement of the susceptibility of representative samples at liquid N temperatures (77 K) indicated that the increase in susceptibility over room temperature values was not great enough to indicate that all the susceptibility was carried by paramagnetic minerals, but paramagnetic minerals did make significant contributions to the susceptibility.

Gunderson et al. (2012) concluded that the susceptibility was carried by a combination of paramagnetic and ferromagnetic minerals. Since precession tends to affect insolation at low and middle latitudes, and would thus affect monsoon strength and runoff from the nearby continent, it was easy to see how precession could control the delivery of magnetite and detrital clays to the depositional basin. However, obliquity does not strongly affect insolation at the low-to-mid latitudes of the Stirone Formation in the Plio-Pleistocene. Obliquity is a strong control on sea level, due to obliquity's heavy influence on high latitude insolation and hence to changes in global ice volume. Using the magnetostratigraphic tie points (reversal boundaries), Gunderson et al. (2012) showed that susceptibility highs are coincident with obliquity minima (and therefore sea level low stands). Gunderson et al. (2012) speculate that lower sea levels lead to restricted Mediterranean circulation, more anoxia, and perhaps enhanced production of iron sulfides.

Therefore, the encoding envisioned for the Stirone Formation comprises two parts: a simple erosional model for the encoding of precession and a more complicated low sea level, more anoxia, more iron sulfides model for the obliquity signal.

6.3 Arguis Formation, Spanish Pyrenees

This important rock magnetic cyclostratigraphy study (Kodama et al. 2010) was used as an example for astronomical tuning discussed in Chapter 5. Details of the identification of the astronomically forced cycles and the tuning procedure can be found in Chapter 5.

The Arguis Formation is a sequence of Eocene marine marls exposed in the External Sierra of the Jaca Basin of the Spanish Pyrenees. The Arguis Formation pro-deltaic sediments are folded, via salt tectonics, into the Pico de Aguilla anticline. Samples were collected from 800 m of growth strata on the flanks of the anticline to provide a high-resolution age model for timing the growth of the fold. The Arguis Formation coarsens up section in grain size from mudstones at the base- to medium-coarse sandstones at the top of the section.

Reconnaissance magnetostratigraphy had been published by Hogan and Burbank (1996) before the rock magnetic cyclostratigraphic study was conducted, so the researchers had an approximate idea of the sediment accumulation rate to plan the sampling, and particularly to set the sampling interval at about five times per predicted precessional cycle (every 4 kyr). Oriented samples for paleomagnetic measurement were collected near reversal boundaries already located by Hogan and Burbank's (1996) magnetostratigraphy, but at a 3 m stratigraphic spacing, to better resolve the reversal boundary locations. More precisely located reversal boundaries would give more accurate absolute time control for calibrating the cyclostratigraphy. Unoriented cyclostratigraphy samples were collected every 20 cm for the bottom 100 m of section, where lithology and the presence of glauconite suggested a condensed section and a slower than average sediment accumulation rate, at 75 cm spacing for the middle 100–500 m part of the section and at a 1.5 m stratigraphic spacing from 500 to 800 m at the top of the section where the magnetostratigraphy (Hogan & Burbank 1996) and the coarser grained lithology suggested high-sediment accumulation rates.

IRM acquisition measurements indicated that the ferromagnetic minerals carrying the primary paleomagnetism of the rocks were magnetite, a primary magnetic mineral, and Fe sulfides, which are secondary magnetic minerals, formed by reductive diagenesis (Roberts & Weaver 2005). The magnetic mineralogy of the Arguis Formation indicated that ARM would be the best way to measure the rock magnetic cyclostratigraphy, with the caveat that secondary Fe sulfides could be carrying at least some of the ARM.

The refined magnetostratigraphy identified eleven polarity intervals and extended the Hogan and Burbank (1996) magnetostratigraphy 250 m up section. Using biostratigraphy (planktonic foraminifera, benthic foraminifera, and calcareous nanoplankton), the reversal stratigraphy could be correlated to Chrons C18–C16 in the Eocene (from about 40 to 36 Ma) giving an average sediment accumulation rate of 23 cm/kyr to help identify astronomically forced cycles in the cyclostratigraphy.

MTM spectral analysis of the ARM cyclostratigraphy revealed significant spectral peaks with approximately 5 m wavelengths (5.6, 5.3, and 4.7 m) which based on the average sediment accumulation rate would be ~22 kyr in duration and likely to be precession. A significant peak was also observed at 14.6 m (~63.5 kyr) that was not identified as astronomically forced. Strong, significant spectral peaks were also observed at 30.7 m (~133.5 kyr), 53 m (~230.4 kyr), and 82 m (~356.5 m). The first peak was strongly suspected to be short eccentricity and the 82 m peak could be long eccentricity (Figure 6.4).

As explained in Chapter 5, the cyclostratigraphic age model for the Arguis Formation was refined by multiple steps of tuning to account for changes in sediment accumulation rate and unrecognized hiatuses throughout the section. The ARM data series was tuned a total of three times to eccentricity. The series was first filtered with a band-pass filter, centered at 405 kyr, to more accurately

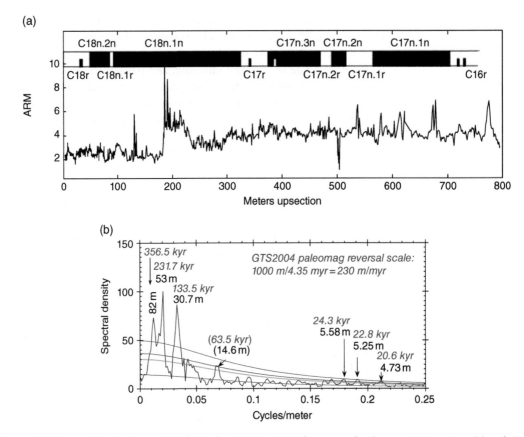

Figure 6.4 Figure from Kodama et al. (2010) showing the ARM data series for the Arguis Formation (a) and its tie to the magnetostratigraphy measured for this rock magnetic cyclostratigraphic study. The MTM power spectrum (b) for the untuned ARM data series tied to time via the magnetostratigraphy with the robust red noise (red) confidence limits of 90% green, 95% blue, and 99% purple shows significant peaks in the precessional band (~5 m) and short (30.7 m) and long (82 m) eccentricities. Source: Kodama, Anastasio, Newton, Pares & L. A. Hinnov 2010. Reproduced with permission of John Wiley & Sons, Inc.

stretch the time series into 405 kyr intervals (tuning 1). Then the minima observed at roughly 100 kyr intervals for the 405 kyr tuned ARM series were matched to 100 kyr minima in the theoretical eccentricity (Laskar et al. 2004) for the Eocene. Finally, the twice-tuned series was filtered (Gaussian band pass) at the 100 kyr scale and the filtered time series was tuned to theoretical eccentricity. The multiple tuning procedure was checked with the MTM power spectra of the series at each tuning step. Basically, the tuning enhanced the 400, 122, and 94 kyr peaks (eccentricity) as expected, but also enhanced precessional peaks at 21.7, 19.1, and 17.4 kyr which matches the theoretical peaks at this time period (23.2, 20.4, and 18.8 kyr) quite well. An additional precessional-scale peak at 26.7 kyr also emerges in the power spectrum that doesn't fit with the theoretical Eocene precessional spectrum suggesting that too much tuning may not be justified. The precessional peaks after two tunings (to long and short eccentricities) resolve into two precessional peaks at 23 and 18.2 kyr that fit the theoretically predicted peaks quite well; however, the non-Milankovitch 178 kyr peak is not eliminated at this level of tuning. Tuning, therefore, should be approached with caution, and the power spectrum should be checked at each step to assess the effects of the tuning.

It is clear from the Arguis Formation study that despite the presence of secondary Fe sulfides in the marine sediments, a successful cyclostratigraphy and a successful magnetostratigraphy can be extracted. These results would suggest that the primary depositional magnetite dominates both the paleomagnetism and the rock magnetic cyclostratigraphy and that the Fe sulfides do not make significant contributions to the rock magnetic cyclostratigraphy.

Kodama et al. (2010) get at the encoding mechanism in a unique way. They conduct a coherency analysis that shows the eccentricity-tuned ARM data series has precession that is in phase with October–November insolation for the Arguis's paleolatitude of 35°N. It is important to note that the Arguis Formation ARM data were not tuned at the precessional scale. ARM precession in phase with October to November insolation would coincide with a Mediterranean-climate's fall rainy season, hence the encoding mechanism is thought to be driven by runoff delivering continentally derived sediment into a relatively constant production of marine carbonate. ARM peaks would occur during greater runoff from the continent.

6.4   Cupido Formation Platform Carbonates, Northeastern Mexico

The rock magnetic cyclostratigraphic study of the Cretaceous (Barremian–Albian) Cupido Formation platform carbonates is an important study because it demonstrates that rock magnetism can detect astronomically forced climate cycles while a repeating sequence of upward shallowing, peritidal cycles does not (Hinnov et al. 2013). The Cupido Formation is located in the Sierra Madre Oriental fold belt of northeastern Mexico. It is 940 m thick and is characterized by meter scale (1–8 m, average thickness: 3.4 m), upward shallowing cycles including subtidal, intertidal, and supratidal facies. The uppermost 150 m of the formation is characterized by facies patterns that indicate a gradual deepening of the depositional environment. The upward shallowing cycles in the Cupido Formation indicate sea level fluctuations that could be due either to processes internal to the depositional basin or to a driving force external to the basin, possibly astronomically forced climate change. Goldhammer et al. (1991) interpreted the upward shallowing cycles to represent obliquity based on biostratigraphic age control. Unfortunately, the biostratigraphic age controls are not robust. At two different localities, the base of the overlying La Pena Formation could either be older than the Oceanic Anoxic Event 1a (OAE1a or "Selli" event, (Schlanger & Jenkyns 1976)) based on planktonic foraminifera or younger than OAE1a based on ammonites. This ambiguity in the age of the top of the Cupido Formation would suggest that the biostratigraphic age controls are not good enough to assign durations to the upward shallowing sequences.

Hinnov et al. (2013) sampled 143 m of the topmost part of the Cupido Formation at two different localities separated by about 25 km. At Garcia Canyon, inner-shelf facies were sampled and at Chico Canyon, middle-shelf facies were collected. Unoriented samples were taken every 20–50 cm, each carbonate upward shallowing cycle was sampled at least four to five times. The sampling was designed to be at a finer scale than the facies to assess if the cyclicity observed by the rock magnetic measurements was independent of the upward shallowing cycles. Two hundred and ninety-seven samples were collected at Garcia and 283 samples at Chico. Samples were crushed into 2–4 mm size pieces and placed in $2 \times 2 \times 2$ cm plastic sample boxes for magnetic measurement. The boxes were completely filled in order to immobilize the rock pieces during measurement. All magnetic measurements were normalized by mass.

Rock magnetic tests in addition to the cyclostratigraphy rock magnetic parameters were made in order to assess the magnetic mineralogy. Low temperature and hysteresis measurements indicate that the magnetic carrier of the cyclostratigraphy is pseudo single domain (PSD) to multidomain (MD) (grain size 0.1–1 μm) magnetite. Based on these results, ARM was used for the rock magnetic cyclostratigraphy. ARM was applied with a 100 mT peak alternating field in the presence of a 97 μT bias field.

The $\log_{10}$ARM cyclostratigraphy shows strong correlation between the inner shelf and mid-shelf localities (Figure 6.5). The $\log_{10}$ of the ARM helps to stabilize the variance making the cycles more obvious. The strongest feature in the

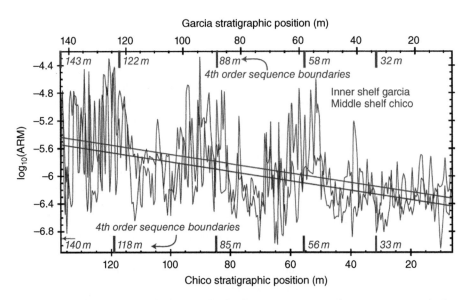

Figure 6.5 $\log_{10}$ARM rock magnetic cyclostratigraphy for the Cretaceous Cupido Formation. Results from two different localities (Garcia and Chico Canyons) show strong correlation of 30–35 m cycles with higher frequency cycles superimposed. Source: Hinnov, Kodama, Anastasio, Elrick & Latta 2013. Copyright 2013 by the Geological Society of London.

cyclostratigraphy is a long wavelength cycle with a period of about 30–35 m that is similar in length to the spacing of fourth order sequence stratigraphy boundaries recognized in the Cupido rocks (Goldhammer 1999). These long-period cycles can be easily matched between Chico and Garcia and are seen to diminish in amplitude up section. The decrease in amplitude may be due to increasing water depth suggested by the facies patterns. Higher frequency variability is superimposed on the dominant 30–35 m cycle. No magneto-stratigraphy was measured for the Cupido cyclostratigraphic study. In lieu of direct measurement of absolute time, identification of orbitally forced cycles may be accomplished by tying one period to a given Milankovitch cycle to see if other astronomically forced cycles are enhanced in the power spectrum. This approach was used for the $\log_{10}$ARM data series in the Cupido Formation. The fourth order sequence boundaries were assumed to be the 405 kyr long eccentricity period. This assumption is supported by tying the two oldest and youngest fourth order sequence boundaries to the Barr6, AP1, and AP2 global sequences suggesting the section sampled is about 1.62 myr in duration. When this tie to time is made, the MTM spectral analysis shows that spectral peaks emerge for the $\log_{10}$ARM data series with periods of 100, 44, and 20 kyr that can be easily tied to short eccentricity, obliquity, and precession (Figure 6.6). This support for astronomically forced rock magnetic cycles is bolstered by the

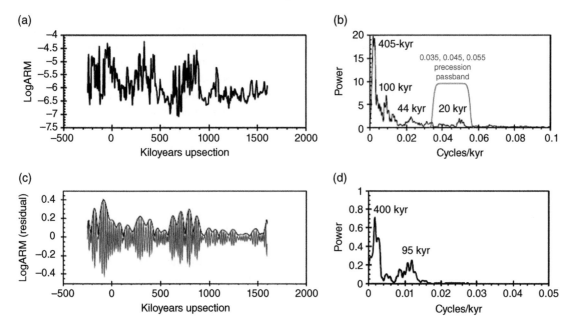

Figure 6.6 Figure from Hinnov et al. (2013) showing $\log_{10}$ARM data series for Garcia (a) and its power spectrum (b) after the fourth order sequence boundaries have been tied to 405 kyr long eccentricity. Astronomically forced cycles emerge at 100 kyr (eccentricity), 44 kyr (obliquity), and 20 kyr (precession). Further testing of the astronomically forced nature of these cycles shows that the amplitude envelope (c) of the $\log_{10}$ARM series filtered at the precessional scale shows (d) spectral peaks of the 400 kyr period of long eccentricity (expected) and the ~100 kyr period of short eccentricity. Source: Hinnov, Kodama, Anastasio, Elrick & Latta 2013. Copyright 2013 by the Geological Society of London.

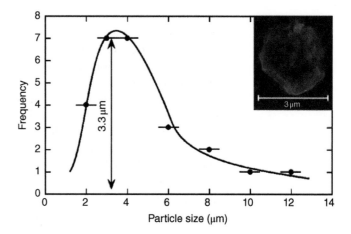

Figure 6.7 Figure from Latta (2005) showing a magnetite grain analyzed by SEM/EDX in upper right corner. The grain is iron oxide with quartz coatings suggesting quartz cement and a detrital origin for the grain. Graph shows grain size measurements and average grain size of 3.3 ± 1 μm that is consistent with that of far-traveled atmospheric dust particles.

observation that the amplitude envelope of the $\log_{10}$ARM series filtered at the precessional scale (Figure 6.6) shows evidence for not only the long eccentricity period (~400 kyr), as expected based on the tuning, but also short eccentricity.

What is most interesting for the Cupido Formation is that the rock magnetic measurements clearly encode orbitally forced climate cycles, but the upward shallowing facies cycles do not. The lithologic cycles were quantified by Hinnov et al. (2013) for spectral analysis by using the standard technique of assigning "facies ranks" for depositional depth. The data sequence for this rank series is tied to time in the same manner as the rock magnetic data series, by assigning the fourth order sequence boundaries to indicate 405 kyr long eccentricity. While Milankovitch periodicities were observed in the rock magnetic parameters, the rank series saw spectral peaks emerge with periods of 180, 72, and 58 kyr, not orbitally forced periods, indicating that the lithologic variations are not driven by Milankovitch global climate cycles.

The suggested encoding mechanism for the Cupido Formation is based on observations that Latta et al. (2006) made on magnetic separates for these rocks. Diana Latta examined the magnetic extracts under an SEM and saw quartz coatings of the original quartz cement on the magnetite grains suggesting that they are detrital in origin (Figure 6.7). Furthermore, a grain size analysis based on visual examination of the extract indicated a median grain size of 3.3 μm. The SEM observed grain sizes are consistent with the hysteresis parameters indicating MD and PSD magnetite grains. Latta et al. (2006) interpreted this information to indicate an eolian origin for the magnetite grains for this platform carbonate depositional environment. Based on the paleolatitude for the site (15°N) and paleogeographic reconstructions for this time in the Cretaceous, Hinnov et al. (2013) and Latta et al. (2006) both came to the conclusion that aridity driven by variations in monsoon strength were controlling the supply of dust to the depositional basin. The source of the dust was most likely northern Africa given the expected global wind circulation for the Cupido's paleolatitude.

6.5 Latemar Massif, Triassic Carbonates, Northern Italy

The Triassic Latemar massif platform carbonates have been studied extensively as an example of an astronomically forced lithologic cyclostratigraphy. Goldhammer et al. (1987) recognized meter-scale upward shallowing peritidal cycles that are bundled at a 5:1 ratio suggesting precession being modulated by short eccentricity. A controversy erupted when the predicted duration for the deposition of the ~600 m thick sequence (~12 myr), assuming precession drove the meter scale upward shallowing cycles, deviated significantly from the duration of deposition based on tightly clustered U-Pb ages for zircons in ash beds distributed throughout the sequence (~1–2 myr) (Mundil et al. 2003). Therefore, the Latemar rocks were a perfect target for rock magnetic cyclostratigraphy given the results from northeastern Mexico that suggested rock magnetic measurements could give a cyclostratigraphy independent of lithologically determined upward shallowing sequences for a platform carbonate.

Kodama and Hinnov (2007) conducted a preliminary rock magnetic cyclostratigraphy study of 40 m of the Latemar rocks exposed at Forecellone. Two hundred samples were collected at a 20 cm stratigraphic spacing. Depth rank analysis of the shallowing upward facies shows a strong spectral peak with a period of 118 cm (Figure 6.8), consistent with the observation of strong lithologic cyclicity at the meter scale at Forcellone. The depth rank analysis also observed a peak at 500 cm, showing the 5:1 bundling that characterizes the Latemar's lithologic cycles.

IRM acquisition measurements showed that low coercivity magnetite (~50 mT mean coercivity) appeared to be the main magnetic mineral with small (<10%) amounts of a higher coercivity mineral (~200 mT) that was identified as hematite (Kodama & Hinnov 2007). The MTM spectral analysis of the MS, ARM, saturation IRM (SIRM), ARM/SIRM, and S-ratio data series all support the depth rank power spectra and the recognition of meter scale lithologic cycles bundled at 5:1. ARM, in particular, shows spectral peaks at 110 and 500 cm, as does SIRM, showing that magnetite concentration variations easily pick up the observed lithologic cycles.

Alessandro Marangon completed a PhD dissertation at the University of Padua studying, in part, the rock magnetic cyclostratigraphy of the Latemar (Marangon 2011). He collected a longer section (100 m) of the Latemar rocks at 20 cm spacing with 484 unoriented samples. His IRM acquisition work supported what was observed by Kodama and Hinnov (2007). Low-coercivity magnetite (~30 mT mean coercivity) appeared to be the dominant magnetic mineral in the rocks, with small amounts (~10%) of higher coercivity hematite. MTM spectral analysis of depth rank, ARM, $\log_{10}$ARM/MS, SIRM data series all showed essentially the same thing: strong spectral peaks at ~1 and 5 m.

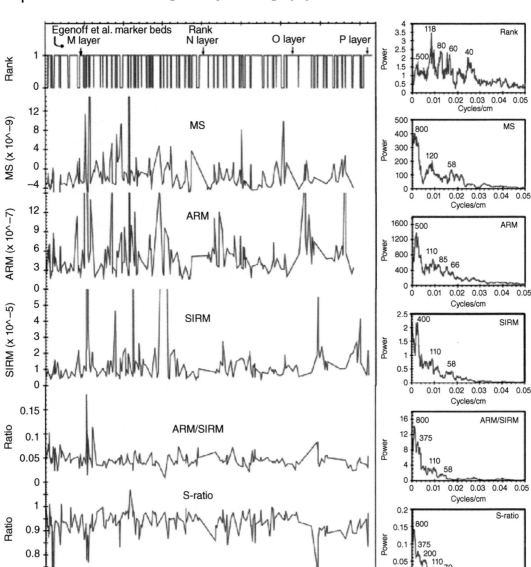

Figure 6.8 Rock magnetic cyclostratigraphy from the Forcellone section of the Latemar. Spectral analysis of depth rank (a) and magnetic parameters (b) show spectral peaks at about 100 and 500 cm. Source: Spahn, Kodama & Preto 2013. Reproduced with permission of John Wiley & Sons, Inc.

The case for assigning the meter scale cycles to precession in the Triassic is based primarily on the 5:1 bundling of the shallowing upward sequences. An independent age control using magnetostratigraphy was attempted by Kent et al. (2004), but the paleomagnetic data suffered from secondary overprinting, most likely caused by lightning strikes. The exposed high elevation outcrops of the Latemar would be a very likely spot for lightning strikes and the high natural remanent magnetization (NRM) intensities that Kent et al. (2004) observed are certainly consistent with IRMs acquired naturally due to the high currents caused by lightning. Kent et al.'s (2004) data showed only one polarity, after samples were excluded because of their high NRM intensities, for nearly all of the section. One polarity interval in the early Triassic would suggest much less than one million years for the duration of the entire Latemar sequence.

Spahn et al. (2013) studied the magnetostratigraphy at Rio Sacuz, a section correlative to the Latemar deposited in the basin adjacent to the Latemar platform. Rio Sacuz was selected because it was below the tree line, less exposed, and less likely to be affected by lightning strikes. In fact, the paleomagnetic data had none of the high NRM intensities observed by Kent et al. (2004). A N-R-N-R polarity sequence was obtained (Figure 3.4) that when correlated to the Triassic GPTS (Hounslow & Muttoni 2010) indicated a one million year duration for the deposition of the Rio Sacuz section, and by correlation, the entire 600 m Latemar sequence. Spahn et al. (2013) also measured the MS throughout the 70 m Rio Sacuz section on the paleomagnetic cores collected every meter of section for the magnetostratigraphic study. MTM spectral analysis revealed significant spectral peaks at 9.8 and 7.9 m which appeared to be modulated by a longer ~25 m cycle which repeated a little more than twice throughout the section (Figure 6.9). The less than significant 23 m spectral peak is likely weak due to only two repetitions. The ~7–9 m peaks are interpreted by Spahn et al. (2013) to be of short eccentricity (125 and 95 kyr) and the 25 m peak is interpreted to be of long eccentricity (405 kyr). Based on the MS cyclostratigraphy, the duration of the Rio Sacuz section is more precisely determined to be ~800–900 kyr. These results suggest that the meter scale shallowing upward cycles in the Latermar are sub-Milankovitch in duration with a period of nominally 1700 years. The Triassic Daye Formation from China (see Section 6.6) also observed sub-Milankovitch cycles of similar duration.

6.6   Daye Formation, Triassic Carbonates, South China

Wu et al. (2012) studied the rock magnetic cyclostratigraphy of the 57 m thick early Triassic Daye Formation. They collected an amazing 2440 specimens, sampling the marine limestones, marls, and shales every 1–3 cm for 55.1 m of section. They crushed the samples and placed them in the standard $2 \times 2 \times 2$ cm plastic boxes used in paleomagnetic studies and mass normalized all their remanence measurements.

Rock magnetic measurements identified low coercivity magnetite (~55 mT) with some evidence of a higher coercivity magnetic mineral, probably hematite. This magnetic mineralogy suite is very similar to what

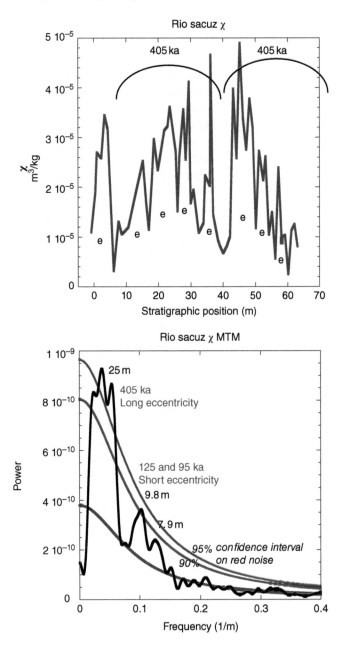

Figure 6.9 Magnetic susceptibility cyclostratigraphy for the Rio Sacuz section. Samples were collected every meter and showed what was interpreted by Spahn et al. [2013] to be long (405 kyr) and short (125 and 95 kyr) eccentricity.

was observed for the Triassic carbonates of the Latemar. Based on the magnetic mineralogy, ARM, applied with a peak alternating field of 100 mT and a bias field of 50 μT, was used for the rock magnetic cyclostratigraphy. The authors point out that the high coercivity hematite, if present in a sample, would not be activated by the 100 mT peak alternating field used to apply the ARM. Magnetic susceptibility was also measured for each sample. Interestingly, even for these weakly magnetized carbonates, the ARM could be measured with a spinner magnetometer (Agico JR6), a relatively inexpensive instrument to purchase and maintain compared to the superconducting rock magnetometers often used for remanence measurements.

Wu et al. (2012) used the log of ARM to stabilize the variance of the ARM data series. They prewhitened their data and used both MTM spectral analysis and wavelet analysis to search for any periodicities. In order to identify astronomically forced cycles, they determined the average sediment accumulation rate for the Daye Formation from an estimate of the duration of the Induan stage in the Triassic to be between 1.0 and 1.4 million years. The upper and lower boundaries of the Induan stage were identified by conodont biostratigraphy.

Many cycles were observed in the ARM and susceptibility power spectra, with long and short eccentricity (13–16 m and 3–4 m), obliquity (1.3–1.7 m), precession (0.69–0.78 m) identified based on the average sediment accumulation rate, as well as many sub-Milankovitch frequency cycles (Figure 6.10).

To further nail down the identification of astronomically forced cycles in the Daye Formation, Wu et al. (2012) conducted amplitude modulation (AM) analysis, similar to that conducted by Hinnov et al. (2013) for the Cretaceous Cupido Formation. Wu et al. (2012) filtered the data series at the precessional scale with a Gaussian band-pass filter and then ran MTM spectral analysis of the envelope of the filtered precession. They observed eccentricity peaks for the power spectrum of the envelope showing that eccentricity modulated precession, exactly what would be expected if the astronomically forced cycles were correctly identified (Figure 6.11). Wu et al. (2012) also performed this analysis on sub-Milankovitch cycles, filtering in the sub-Milankovitch frequency (4–5 kyr periods) and detecting Milankovitch frequencies in the MTM spectrum of the envelope (precession, obliquity, and eccentricity), suggesting that orbital frequencies modulate the sub-Milankovitch frequencies.

Wu et al. (2012) suggest an encoding mechanism for the Daye Formation similar to that envisioned for the Eocene Arguis Formation (Kodama et al. 2010). The variations in the concentration of fine-grained, low-coercivity magnetite that carries the orbital and suborbital frequency periodicities are due to the fluctuations of siliciclastic terrestrial input into the platform carbonates of the Daye Formation. Wu et al. (2012) suggest that these fluctuations are due to changes in the strength of the monsoon at the low paleolatitudes of the Daye Formation. The strong modulation of precession by eccentricity in the ARM and MS records is the primary basis for this interpretation since is it recognized that at low latitudes precession controls the strength of the monsoon. Hence, according to Wu et al. (2012), eccentricity maximum and precession minima caused higher continental runoff and, consequently, greater magnetite concentration.

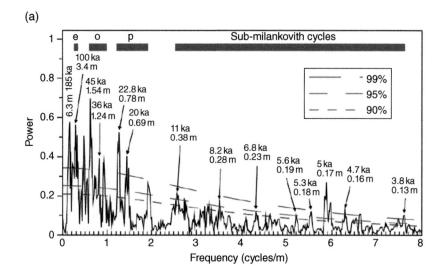

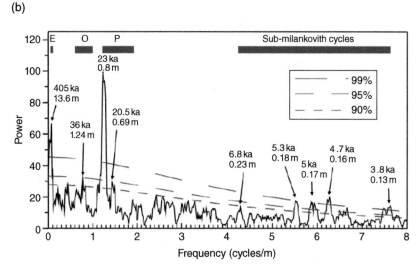

Figure 6.10 (a) Spectral analysis of ARM (b) and MS data series from the Triassic Daye Formation carbonates. The duration of the different cycles is based on assuming that the biostratigraphically identified Induan stage in the Triassic was between 1 and 1.4 million years in duration. Note the identification of both astronomically forced cycles and sub-Milankovitch cycles. Source: Wu et al. 2012. Reproduced with permission of Elsevier.

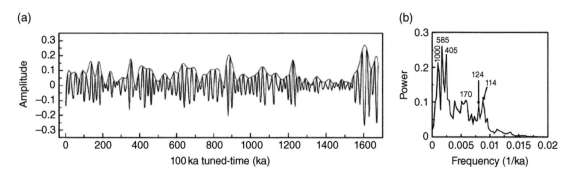

Figure 6.11 AM (amplitude modulation) of the ARM series from the (a) Daye Formation showing that the envelope (in red) of the precessional signal (obtained by Gaussian band-pass filtering) has an (b) MTM spectrum showing cycles with periods of short and long eccentricity, showing that precession is modulated by eccentricity. Source: Wu et al. 2012. Reproduced with permission of Elsevier.

6.7   Mauch Chunk Formation: Mississippian
Red Beds, Pottsville, Pennsylvania

The magnetostratigraphy of the upper Mississippian (upper Visean ~328 Ma) red beds of the Mauch Chunk Formation has been studied extensively (DiVenere & Opdyke 1991). These rocks were, therefore, a natural target for a rock magnetic cyclostratigraphy study, since absolute time has already been established for them. One of the goals of this cyclostratigraphic study was to see if astronomically forced cycles can be detected rock magnetically in the Paleozoic. Another goal was to determine if terrestrial rocks deposited in a fluvial environment can record Milankovitch-scale climate cycles. As Jerolmack and Paola (2010) pointed out, a fluvial system has the strong potential for shredding environmental signals because of depositional and transport processes. Two important factors that would limit the ability of a fluvial depositional environment's ability to record climate cycles are the discontinuous deposition in a river's floodplain and the effects of channel migration across the floodplain.

A well-studied stratigraphic section of the Mauch Chunk Formation was sampled near Pottsville, Pennsylvania. Levine and Slingerland (1987) interpreted this section to show that the Mauch Chunk was deposited in a braided, alluvial plain in a semi-arid environment subject to catastrophic floods. The Pottsville site has good stratigraphic control based on Levine and Slingerland's (1987) detailed stratigraphic section and has 2–3 m thick channel deposits with 20–30 m thick sections of floodplain (overbank) deposits. DiVenere and Opdyke (1991) conducted a magnetostratigraphic study of the Pottsville section and their results suggest that the stratigraphy is continuous for about 1.2 myr.

A reconnaissance rock magnetic cyclostratigraphy study was conducted using a hand-held, portable susceptibility meter (GF Instruments, model SM20). A measurement was made by simply taking a background measurement in the air and then placing the hand-held meter on the rock outcrop for the rock susceptibility measurement. Susceptibility measurements were collected over 63 m of section at a 50 cm sampling interval in only several hours. Samples did not need to be collected from the outcrop or taken back to the laboratory for preparation and measurement. The data could be processed almost immediately. The susceptibility data set showed strong, hierarchical cyclicity with meter-scale high-frequency cycles superimposed on longer ~10 m cycles (Figure 6.12).

MTM spectral analysis of the susceptibility data series shows three significant spectral peaks (>99% confidence level) with wavelengths of 13.3, 10.3, and 1.7 m, and peaks that rise about the 95% confidence level at 2.6 and 2.1 m wavelength (Figure 6.13).

DiVenere and Opdyke (1991) recognize a normal polarity interval that is 100 m thick in the lower part of the Pottsville outcrop very close stratigraphically to where the susceptibility readings were made. According to Opdyke

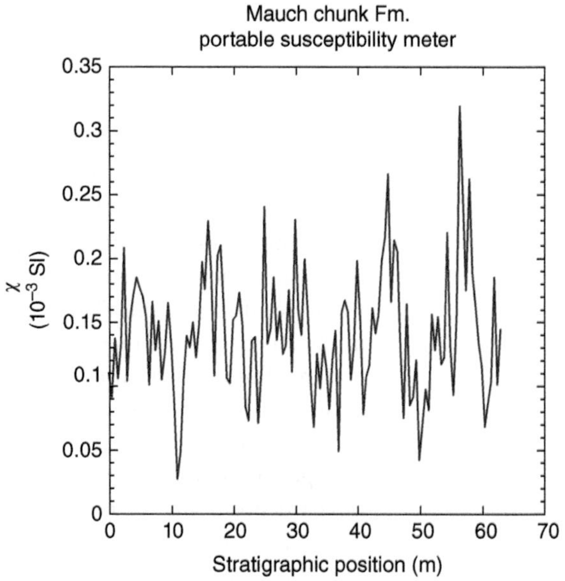

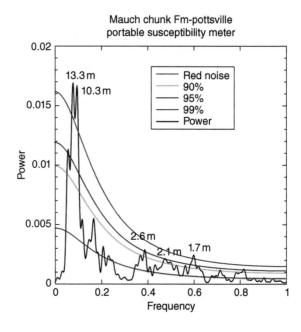

Figure 6.12 Magnetic susceptibility (χ) for 63 m of the Mauch Chunk Formation collected in 0.5 m intervals at Pottsville, Pennsylvania. Note the meter scale variability superimposed on ~10 m cycles.

Figure 6.13 MTM spectral analysis of susceptibility rock magnetic cyclostratigraphy collected with a hand-held susceptibility meter for the Mauch Chunk Formation at Pottsville, Pennsylvania. If the 11 cm/kyr sediment accumulation rate determined by magnetostratigraphy is applied to the spectral peaks, 13.3 and 10.3 m are short eccentricities and the 2.6, 2.1, and 1.7 m peaks are in the precessional band.

et al.'s (2000) calibration to the Carboniferous and the most recent Gradstein et al. (2012) geologic timescale, this normal polarity interval is 890 kyr in duration that would suggest an 11 cm/kyr sediment accumulation rate for the Mauch Chunk Formation at Pottsville. The magnetostratigraphically constrained sediment accumulation rate would indicate that the 13.3 m spectral peak is 121 kyr in duration, the 10.3 m peak is 94 kyr in duration, and the 1.7 m peak is 16 kyr in duration. The less significant peaks at 2.6 and 2.1 m would indicate 24 and 19 kyr, respectively.

Laskar et al. (2011) indicate that eccentricity has not changed significantly in duration throughout geologic time. Based on the magnetostratigraphy, the 13.3 and 10.3 m cycle are very close to the expected duration of the short eccentricity peaks (125 and 95 kyr, see Chapter 5). The 1.7 m (16 kyr), 2.6 m (24 kyr), and 2.1 m (19 kyr) peaks are reasonably close to Berger and Loutre's (1994) estimate for precession band peaks at 298 Ma (20.7 and 17.4 kyr). From the simple reconnaissance rock magnetic cyclostratigraphy study of susceptibility, it appears that astronomically forced cycles can be recorded in a fluvial environment, despite the potential for shredding caused by depositional and erosional processes. The results from the hand-held susceptibility meter show, too, that reconnaissance studies can be easily done to check if a sedimentary sequence is suitable for rock magnetic cyclostratigraphy and what the appropriate sampling intervals should be for an intensive cyclostratigraphy study with remanence measurements.

In a separate rock magnetic study of the Pottsville, Pennsylvania, outcrop of the Mauch Chunk Formation, both isothermal remanent magnetization (IRM) and susceptibility measurements were made in the laboratory on samples collected every 50 cm for 40 m of stratigraphic section. These results showed that the laboratory remanence and susceptibility measurements both detected the eccentricity cycles, suggesting ferromagnetic minerals dominate the susceptibility measurements. If this is true, it indicates that the portable susceptibility measurements detect variations in hematite concentration that appear to record astronomically forced climate change.

6.8 Rainstorm Member of the Neoproterozoic Johnnie Formation, Death Valley, California

The cyclostratigraphy and magnetostratigraphy of the Rainstorm Member of the Neoproterozoic Johnnie Formation was studied by Minguez et al. (2014) and Kodama and Hillhouse (2011). The Rainstorm Member of the Johnnie Formation is comprised of gray, green, pink, and purple marine siltstones and carbonates exposed near Death Valley, California. It is made up of three units: a basal siltstone, a middle carbonate unit, and an upper unit of interbedded siltstone and finely bedded quartzite. Near the middle of the basal greenish gray to purplish siltstone unit is a distinctive ~2 m thick oolitic layer, the Johnnie Oolite, which serves as an important marker bed used for regional correlation. The thickness of the Rainstorm Member varies significantly in the Death Valley region. The cyclostratigraphy and magnetostratigraphy studies were conducted at two localities, the Nopah Range and Winters Pass Hills, and at these two localities, the Rainstorm Member is 120 and 60 m thick, respectively (Verdel et al. 2011).

The goal of the cyclostratigraphy study was to determine the duration of the Shuram carbon isotope excursion that has been found in the Rainstorm Member (Corsetti & Kaufman 2003). The Shuram Excursion is a significant negative excursion in $\delta^{13}C$ isotope values, in which $\delta^{13}C$ values plummet from +5 to as low as −12‰ and recover slowly over millions to perhaps tens of millions of years. It is observed globally, in South China, Oman, South Australia, Namibia, India, Mexico, and Death Valley, and it is thought to record the oxidation of organic carbon in an anoxic Neoproterozoic ocean just before the explosion of multicellular life in the Cambrian (Kaufman et al. 2007; McFadden et al. 2008; Grotzinger et al. 2011). The Shuram Excursion is observed stratigraphically for several hundred meters, starting just above the Johnnie Oolite (Verdel et al. 2011), although the excursion is not complete at the Nopah Range and Winters Pass Hills localities because of a postdepositional incision event (Corsetti & Kaufman 2003). The depositional environment of the Rainstorm Member was a shallow marine continental shelf and was apparently affected by strong storm events as evidenced by thin siltstone units interbedded into carbonate (Summa 1993; Pruss et al. 2008).

About 40 m of the Rainstorm Member was sampled in the Nopah Range, and about 55 m was sampled in the Winters Pass Hills. Unoriented samples were collected every 20–25 cm stratigraphically. Oriented samples were also collected for paleomagnetic measurement to develop a magnetostratigraphy to constrain the sediment accumulation rate for these two localities. Nine horizons were collected every 4–6 m in the Nopah Range and 19 horizons every 3-4 m in the Winters Pass Hills.

The Nopah Range locality gave better paleomagnetic results and a mean paleomagnetic direction (D = 262.8°, I = 1.3°, α_{95} = 16.4°) similar to that observed by Van Alstine and Gillett (1979) for the Rainstorm Member in the Desert Range of Nevada about 150 km to the northeast (D = 258°, I = −1°). At the Nopah Range, an R-N-R-N magnetostratigraphy was observed over the 40 m sampled. The paleomagnetic data from the Winters Pass Hills locality suffered from extensive overprinting, probably a Cretaceous age viscous thermal partial remagnetization in a north and down direction, on the east (reversed)-west (normal) and shallow directions from the Neoproterozoic. The overprinting makes it difficult to determine polarity intervals, particularly because the directions are not antipodal, but there is clearly a long normal polarity interval of about 30 m in stratigraphic thickness with a 7 m thick reversed polarity interval stratigraphically below and mixed polarity layers above and below this sequence. The long normal polarity interval at the Winters Pass Hills can be correlated with the ~20 m thick lowermost normal polarity interval at the Nopah Range. Thermal demagnetization experiments show that almost all of the remanence is removed at temperatures near to 580°C, indicating magnetite as the paleomagnetic carrier. Magnetite is most likely a primary, depositional magnetic mineral suggesting that the remanence is primary. The number of polarity intervals in these sections can only provide an order of magnitude estimate for the duration of their deposition. Assuming a maximum geomagnetic field, reversal rate similar to that of the last 160 million years (Merrill et al. 1996) occurred during the Proterozoic (Pavlov & Gallet 2010); a minimum duration for deposition of the sediment at these localities would be somewhat less than one million years.

Magnetic susceptibility, ARM, and IRM were measured on the unoriented samples to determine the magnetic mineral concentration variations throughout the two sections and construct cyclostratigraphy data series for time series analysis. ARM was applied with a 100 mT peak alternating field in the presence of a 97 μT bias field. The IRM was acquired in a 5 T field. Since magnetite appears to be carrying the paleomagnetism and the successful observation of a magnetostratigraphy that can be correlated between the two localities suggests it is a primary magnetization, this bodes well for the cyclostratigraphy, particularly the ARM data series which targets the magnetite in the rocks. Kodama and Hillhouse (2011) also constructed a magnetic parameter ratio to measure the goethite to hematite ratio in the rocks. The goethite to hematite ratio of a sedimentary sequence is used to detect changes in precipitation or moisture availability in the source area (e.g., Harris & Mix 2002), a magnetic goethite to hematite ratio was measured as a possible way to detect climate variability. Goethite was measured by the intensity loss from thermal demagnetization of a 4T IRM at 130°C. Hematite was determined from the 4T IRM, after goethite had been removed by thermal demagnetization and magnetite by alternating field demagnetization (or subtraction of the ARM applied in a 100 mT alternating field). The goethite/hematite (G/H) ratio can be written as:

$$\frac{G}{H} = \frac{\left[IRM_{4T} - IRM_{4T\,130\,°C} \right]}{IRM_{4T\,130\,°C\,AF100\,mT}} \tag{6.1}$$

where IRM_{4T} is an IRM applied in a 4 T field, $IRM_{4T\,130°C}$ is the 4 T IRM thermally demagnetized at 130°C, and IRM $_{4T\,130°C\,AF100mT}$ is the 4 T IRM that has been thermally demagnetized and alternating field demagnetized in a 100 mT peak alternating field. Significant spectral peaks in the MTM power spectra of the MS, ARM, and IRM data series with periods of 16–18 m, 5–6 m, 1–2 m, and 0.5–0.6 m were observed at both the Nopah Range and Winters Pass Hills localities. Based on the approximately one million year duration for the deposition of the sections, this would indicate that the 16–18 m spectral peak probably represents the 405 kyr long eccentricity which according to Laskar et al. (2011) should not vary significantly through geologic time. This time assignment is supported by the observation that if 18 m represents 405 kyr of time, then 5 m is ~110 kyr and is probably short eccentricity, 1–2 m is ~34 kyr which is most likely obliquity, and 0.6 m is 13.5 kyr which is probably precession. Berger and Loutre (1994) estimate the periods for the Milankovitch cycles at different times in the geologic past,

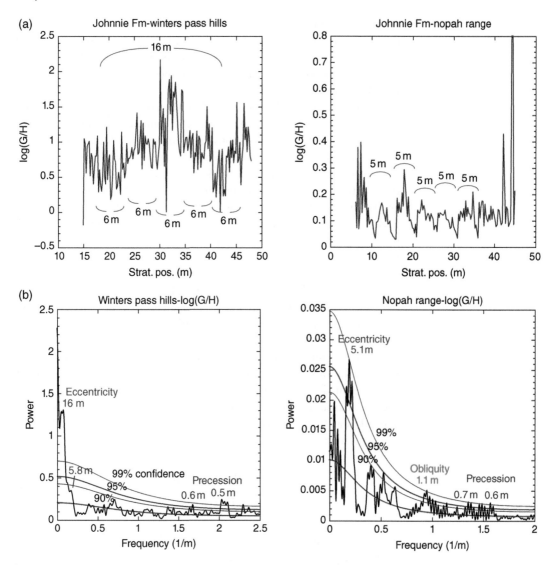

Figure 6.14 $\log_{10}$(G/H = goethite to hematite) as a function of stratigraphic position for the Winters Pass Hills and Nopah Range localities (a). Both the 5–6 m and 16 m wavelengths are observable in the data series. MTM spectral analysis of the $\log_{10}$(G/H) data series shows significant peaks at 16, 5, 1.1, and 0.6 m for the $\log_{10}$(G/H) series (b).

fully cognizant that the evolution of the Earth–Moon system is poorly constrained and thus obliquity and precession cannot be accurately calculated for deep geologic time. However, Berger and Loutre (1994) indicate that obliquity was approximately 29.9 kyr in duration and precessional periods were 19 and 16.2 kyr long at 500 Ma.

Strong 5 m cycles and 0.6 m cycles, as well as 16 and 1 m cycles, in the $\log_{10}$ of the goethite to hematite magnetic ratio (Figure 6.14) were also observed at both the Nopah Range and Winters Pass Hills. The 5 and 0.6 m cycles are probably short eccentricity and precession, while 16 and 1.1 m peaks are most likely long eccentricity and obliquity suggesting the G/H magnetic parameter may be a viable way of detecting ancient climate cycles magnetically. However, it is probably incorrect to interpret the goethite to be depositional in the Rainstorm Member, and hence a measure of moisture in its Neoproterozoic source area. Thermal demagnetization suggests that the goethite, removed at low temperatures in thermal demagnetization, carries a present-day field direction indicating that it probably formed during recent surface weathering. Furthermore, conodont alteration indices of 4.5 in Ordovician rocks stratigraphically above the Johnnie Formation (Gillett 1982) indicate that the Johnnie Formation was heated at least to temperatures of 200–300°C which would have been hot enough to invert any ancient goethite to hematite (deFaria & Lopes 2007). It is more likely that present-day weathering converted Fe-rich silicates or clays to goethite. Therefore, $\log_{10}$(G/H) measures subtle variations in Fe-rich clays in the rocks, and hence fluctuations in the input of terrestrial sediment into the carbonate.

The encoding of the astronomically forced climate cycles by the magnetics is similar to that suggested for the Triassic Daye Formation from China (Wu et al. 2012) and the Eocene Arguis Formation marine marls (Kodama et al. 2010). The magnetite concentration probably records variations in terrigenous sediment input into marine carbonate being produced at a relatively constant rate. This interpretation suggests that magnetic methods can be a sensitive way of detecting variations in the amount of continental sedimentation into a near shore marine environment.

6.9 Encoding of Orbitally Forced Climate Signals

The case studies presented here indicate that one of the main mechanisms for encoding climate variations by magnetic mineral concentration is the input of varying amounts of terrestrial sediment into a relatively constant background of marine carbonate production. This was the explanation for the encoding of Milankovitch climate cycles by magnetite concentration in the Eocene Arguis Formation (Kodama et al. 2010), the Triassic Daye Formation (Wu et al. 2012), the Neoproterozoic Johnnie Formation, and at least part of the astronomically forced climate signal (precession and eccentricity) detected in the Plio-Pleistocene Stirone River section (Gunderson et al. 2012). The Arguis Formation interpretation is backed up by an interesting cross-correlation analysis that suggests that ARM maxima coincide in phase with autumn insolation and, therefore, the rainy season for low latitude, monsoonal climate. Of course, it is also possible that carbonate production could have varied due to astronomically forced climate change. There is no way of determining whether magnetic mineral concentration variations in the rocks studied are due to changes in terrestrial input or changes in carbonate production. For the Arguis Formation, though, Kodama et al. (2010) argue that the locality was too far north to be affected by equatorial upwelling and hence changes in carbonate production.

The platform carbonate sections studied, the Cretaceous Cupido Formation and the Latemar from the Italian Dolomites, are interpreted to have acquired their magnetite by eolian deposition, and thus the magnetite concentration variations are reacting to changes in global aridity, driven by precession and modulated by eccentricity. This last option for encoding probably works best in depositional environments where continental sedimentation, delivered by rivers, is not important, so that the magnetic particles associated with eolian dust are not overwhelmed by the magnetic particles transported by runoff.

No encoding mechanism has been proposed for the fluvial depositional environment of the Mauch Chunk Formation. For the Mauch Chunk, the susceptibility probably detects variations in antiferromagnetic hematite for these red rocks. Since magnetostratigraphic studies of the Mauch Chunk Formation (DiVenere & Opdyke 1991) show that the hematite is either depositional or a very early secondary magnetic mineral, the hematite could indicate changes in runoff (transport) or, more likely, if the hematite grew in the sediment above the water table during the dry season (Whidden et al. 1998; Kodama 2012), then the growth of hematite could be a sensitive measure of the average length of the dry season at precessional timescales. Gunderson et al. (2012) also suggest an interesting correlation between obliquity-driven sea level change and anoxia in the Mediterranean basin, thus creating conditions for relatively more production of Fe sulfides.

It is clear from the case studies presented that magnetic mineral concentration can encode climate variations. The exact encoding mechanism can only be determined by detailed examination of the rock magnetic mineralogy on a case-by-case basis, but in general, it appears that variations in the delivery of terrigenous sediments, by air or by sea, to a background of relatively nonmagnetic marine carbonate may be the most likely mechanism.

References

Berger, A. & Loutre, M.F. (1994) Astronomical forcing through geological time. *Special Publication—International Association of Sedimentologists*, 19, 15–24. DOI:10.1002/9781444304039.CH2.

Channell, J.E.T., Poli, M.S., Rio, D., Sprovieri, R., & Villa, G. (1994) Magnetic stratigraphy and biostratigraphy of Pliocene 'argilleazzurre' (Northern Apennines, Italy). *Palaeogeography, Palaeoclimatology, Palaeoecology*, 110, 83–102. DOI:10.1016/0031-0182(94)90111-2.

Corsetti, F.A. & Kaufman, A.J. (2003) Stratigraphic investigations of carbon isotope anomalies and Neoproterozoic ice ages in Death Valley, California. *Geological Society of America Bulletin*, 115, 916–932. DOI:10.1130/B25066.1.

deFaria, D.L.A. & Lopes, F.N. (2007) Heated goethite and natural hematite: Can Raman spectroscopy be used to differentiate them?. *Vibrational Spectroscopy*, 45, 117–121. DOI:10.1016/j.vibspec.2007.07.003.

DiVenere, V.J. & Opdyke, N.D. (1991) Magnetic polarity stratigraphy in the uppermost Mississippian Mauch Chunk Formation, Pottsville, Pennsylvania. *Geology*, 19, 127–130. DOI:10.1130/0091-7613(1991)019<0127:MPSITU>2.3.CO;2.

Gillett, S.L. (1982) Remagnetization and tectonic rotation of upper Precambrian and lower Paleozoic strat from the Desert Range, sourthern Nevada. *Journal of Geophysical Research*, 87, 10929–10953. DOI:10.1029/JB087iB13p10929.

Goldhammer, R.K. (1999) Mesozoic sequence stratigraphy and paleogeographic evolution of northeast Mexico. In: Bartolini, C., Wilson, J.L., & Lawton, T.F. (eds), *Mesozoic Sedimentary and Tectonic History of North-Central Mexico*, pp. 1–58, Geological Society of America, Special Papers, Boulder.

Goldhammer, R.K., Dunn, P.A., & Hardie, L.A. (1987) High-frequency glacioeustatic sea level oscillations with Milankovitch characteristics recorded in Middle Triassic platform carbonates in Northern Italy. *American Journal of Science*, 287, 853–892. DOI:10.2475/ajs.287.9.853.

Goldhammer, R.K., Lehmann, P.J., Todd, R.G., Wilson, J.L., Ward, W. C., & Johnson, C.R. (1991) *Sequence Stratigraphy and Cyclostratigraphy of the Mexozoic of the Sierra Madre Oriental, Northeast Mexico, a Field Guidebook*. Gulf Coast Section, Society of Economic Paleontologists and Mineralogists, Tulsa.

Gradstein, F.M., Ogg, J.G., Schmitz, M.D., & Ogg, G. M., eds. (2012) *The Geologic Time Scale 2012*, 1144 pp. Elsevier, Oxford.

Grotzinger, J.P., Fike, D., & Fischer, W. (2011) Enigmatic origin of the largest-known carbon isotope excursion in Earth's history. *Nature Geoscience*, 4, 285–292. DOI:10.1038/ngeo1138.

Gunderson, K.L., Kodama, K.P., Anastasio, D.J., & Pazzaglia, F.J. (2012) Rock-magnetic cyclostratigraphy for the Late Pliocene-Early Pleistocene Stirone section, Northern Apennine mountain front, Italy. *Geological Society, London, Special Publications*, 373, 26. DOI:10.1144/SP373.8.

Harris, S.E. & Mix, A.C. (2002) Climate and tectonic influences on continental erosion of tropical South America 0-13 Ma. *Geology*, 30, 447–450. DOI:10.1130/0091-7613(2002)030<0447:CATIOC>2.0.CO;2.

Hinnov, L.A., Kodama, K.P., Anastasio, D.J., Elrick, M., & Latta, D.K. (2013) Global Milankovitch cycles recorded in rock magnetism of the shallow marine lower Cretaceous Cupido Formation, northeastern Mexico. In: Jovane, L., Herrero-Bervera, E., Hinnov, L.A., & Housen, B.A. (eds), *Magnetic Methods and the Timing of Geological Processes*, 15 pp. Geological Society, London, Special Publications, London.

Hogan, P.J. & Burbank, D.W. (1996) Evolution of the Jaca piggback basin and emergence of the external Sierra, southern Pyrenees. In: Friend, P.F. & Dabrio, C.J. (eds), *Tertiary Basins of Spain the Stratigraphic Record of Crustal Kinematics*, pp. 153–160. Cambridge University Press, New York.

Hounslow, M.W. & Muttoni, G. (2010) The geomagnetic polarity timescale for the Triassic: Linkage to stage boundary definitions. *Geological Society, London, Special Publications*, 334, 61–102. DOI:10.1144/SP334.4.

Jerolmack, D.J. & Paola, C. (2010) Shredding of environmental signals by sediment transport. *Geophysical Research Letters*, 37. DOI:10.1029/2010GL044638.

Kaufman, A.J., Corsetti, F.A., & Varni, M.A. (2007) The effect of rising atmospheric oxygen on carbon and sulfur isotope anomalies in the Neoproterozoic Johnnie

Formation, Death Valley, USA. *Chemical Geology, 237,* 47–63. DOI:10.1016/j.chemgeo.2006.06.023.

Kent, D.V., Muttoni, G., & Brack, P. (2004) Magnetostratigraphic confirmation of a much faster tempo for sea-level change for the Middle Triassic Latemar platform carbonates. *Earth and Planetary Science Letters, 228,* 369–377. DOI:10.1016/j.epsl.2004.10.017.

Kodama, K.P. (2012) *Paleomagnetism of Sedimentary Rocks: Process and Interpretation,* 157 pp. Wiley-Blackwell, Oxford.

Kodama, K.P. & Hillhouse, J.W. (2011) Rock magnetic cyclostratigraphy and magneto-stratigraphy of the Rainstorm Member of the Neoproterozoic Johnnie Formation indicates 2.5 million year duration for the negative C-isotope excursion. In: *Abstract GP53A-07, 2011 Fall Meeting, AGU,* San Francisco, 5–9 December, 2011.

Kodama, K.P. & Hinnov, L.A. (2007) Mineral magnetic parameters provide new evidence on the climate-driver of the Triassic Latemar carbonate cycles. In: *2007 GSA Denver Annual Meeting,* 28–31 October 2007, paper no. 45-3, .

Kodama, K.P., Anastasio, D.J., Newton, M.L., Pares, J., & Hinnov, L.A. (2010) High-resolution rock magnetic cyclostratigraphy in an Eocene flysch, Spanish Pyrenees. *Geochemistry, Geophysics, Geosystems, 11.* DOI:10.1029/2010GC003069.

Laskar, J., Robutel, P., Joutel, F., Gastineau, M., Correia, A.C.M., & Levrard, B. (2004) A long term numerical solution for the insolation quantitties of the Earth. *Astronomy and Astrophysics, 428,* 261–285. DOI:10.1051/0004-6361:20041335.

Laskar, J., Fienga, A., Gastineau, M., & Manche, H. (2011) La2010: A new orbital solution for the long-term motion of the Earth. *Astronomy and Astrophysics, 532 (A89).* DOI:10.1051/0004-6361/201116836.

Latta, D.K. (2005) *Structural, lithotectonic, and rock magnetic studies of decollement folding, Coahuila marginal folded province, northeast Mexico,* PhD thesis, Lehigh University, Bethlehem.

Latta, D.K., Anastasio, D.J., Hinnov, L.A., Elrick, M., & Kodama, K.P. (2006) Magnetic record of Milankovitch rhythms in lithological noncyclic marine carbonates. *Geology, 34,* 29–32. DOI:10.1130/G21918.1.

Levine, J.R. & Slingerland, R. (1987) Upper Mississippian to Middle Pennsylvanian stratigraphic section Pottsville, Pennsylvania. In: *Geological Society of America Centennial Field Guide-Northeastern Section,* pp. 59–63. Geological Society of America, Denver.

Mann, M. & Lees, J. (1996) Robust estimation of background noise and signal detection in climatic time series. *Climate Change, 33,* 409–445. DOI:10.1007/BF00142586.

Marangon, A. (2011) *Stratigraphic analyses on Monte Agnello and Latemar platforms,* PhD thesis, University of Padova, Padua.

Mary, C., Iaccarino, S., Courtillot, V., Besse, J., & Aissaoui, D.M. (1993) Magnetostratigraphy of Pliocene sediments from the Stirone River (Po Valley). *Geophysical Journal International, 112,* 359–380. DOI:10.1111/j.1365-246X.1993.tb01175.x.

McFadden, P.L., Huang, J., Chu, X., Jiang, G., Kaufman, A.J., Zhou, C., Yuan, X., & Xiao, S. (2008) Pulsed oxidation and biological evolution in the Ediacaran Doushantuo Formation. *Proceedings of the National Academic Sciences, 105,* 3197–3202. DOI:10.1073/pnas.0708336105.

Merrill, R.T., McElhinny, M.W., & McFadden, P.L. (1996) *The Magnetic Field of the Earth, Paleomagnetism, the Core, and the Deep Mantle.* Academic Press, San Diego.

Minguez, D.A., Kodama, K.P., & Hillhouse, J.W. (2014) Paleomagnetic and cyclostratigraphic constraints on the duration of the Shuram carbon isotope excursion, Johnnie Formation, Death Valley region, CA, Geochem., Geophys., Geosys. (in review).

Mundil, R., Zuhlke, R., Bechstadt, T., Peterhansel, A., Egenhoff, S.O., Oberli, F., Meirer, M., Brack, P., and Rieber, H. (2003) Cyclicities in Triassic platform carbonates: Synchronizing radio-isotopic and orbital clocks. *Terra Nova, 15*, 81–87. DOI:10.1046/j.1365-3121.2003.00475.x.

Opdyke, N.D., Roberts, J., Claoue-Long, J., Irving, E., & Jones, P.J. (2000) Base of the Kiaman: Its definition and global stratigraphic significance. *Geological Society of America Bulletin, 112*, 1315–1341. DOI:10.1130/0016-7606(2000)1122.0.CO;2.

Pavlov, V. & Gallet, Y. (2010) Variations in geomagnetic reversal frequency during the Earth's middle age. *Geochemistry, Geophysics, Geosystems, 11*. DOI:10.1029/2009GC002583.

Pruss, S.B., Corsetti, F.A., & Fischer, W. (2008) Seafloor-precipitated carbonates fans in the Neoproterozoic Rainstorm Member of the Johnnie Formation, Death Valley Region, USA. *Sedimentary Geology, 207*, 34–40. DOI:10.1016/j.sedgeo.2008.03.005.

Roberts, A.P. & Weaver, R. (2005) Multiple mechanisms of remagnetization involving sedimentary greigite (Fe3S4). *Earth and Planetary Science Letters, 231*, 263–277. DOI:10.1016/j.epsl.2004.11.024.

Schlanger, S.O. & Jenkyns, H.C. (1976) Cretaceous oceanic anoxic events: Causes and consequences. *Geologie en Mijnbouw, 55*, 179–184.

Spahn, Z.P., Kodama, K.P., & Preto, N. (2013) High-resolution estimate for the depositional duration of the Triassic Latemar Platform: A new magnetostratigraphy from basinal sediments at Rio Sacuz, Italy, *Geochemistry, Geophysics, Geosystems, 14*. DOI:10/1002/ggge.20094.

Summa, C.L. (1993) *Sedimentologic, stratigraphic, and tectonic controls of a mixed carbonate-siliclastic succession; Neoproterozoic Johnnie Formation, southeast California*, PhD thesis, Massachusetts Institute of Technology, Cambridge.

Van Alstine, D. & Gillett, S. (1979) Paleomagnetism of the upper Precambrian sedimentary rocks from the Desert Range, Nevada. *Journal of Geophysical Research, 84*, 4490–4500. DOI:10.1029/JB084iB09p04490.

Verdel, C., Wernicke, B.P., & Bowring, S.A. (2011) The Shuram and subsequent Ediacaran carbon isotope excursions from southwest Laurentia, and implications for environmental stability during the metazoan radiation. *Geological Society of America Bulletin, 123*, 1539–1559. DOI:10.1130/B30369.1.

Whidden, K.J., Lund, S.P., & Bottjer, D.J. (1998) Paleomagnetic evidence that the cetnral block of Salinia (California) is not a far-traveled terrane. *Tectonics, 17*, 329–343. DOI:10.1029/97TC03021.

Wu, H., Zhang, S., Feng, Q., Jiang, G., Li, H., & Yang, T. (2012) Milankovitch and sub-Milankovitch cycles of the early Triassic Daye Formation, South China and their geochronological and paleoclimatic implications. *Gondwana Research, 22*, 748–759. DOI:10.1016/j.gr.2011.12.003.

7 Doing Rock Magnetic Cyclostratigraphy

Abstract: This chapter outlines the steps necessary to complete a successful rock magnetic cyclostratigraphy study. Sampling strategy, sampling techniques, sample preparation, rock magnetic measurements, time series analysis, and tuning are described in the context of conducting a cyclostratigraphy study. Identification of astronomically forced cycles, particularly the 405 kyr cycle of long eccentricity, by the assignment of absolute time to the sedimentary sequence by various techniques, concludes the chapter. This chapter provides a detailed, step-by-step guide for conducting a rock magnetic cyclostratigraphy study. Much of the background information necessary for understanding and analyzing rock magnetic and cyclostratigraphy data sets has been provided in Chapters 2–6. The goal of this chapter is to give researchers enough information to conduct a rock magnetic cyclostratigraphy study, to make them cognizant of potential pitfalls, and to point toward important resources for understanding their data. The subsections of this chapter provide a general outline of the steps, in order, of completing a study.

7.1 Study Design

Before any fieldwork is conducted, important information needs to be collected to plan the study and to determine whether the sedimentary sequence being targeted is likely to give good results. Not all sedimentary rock types are amenable to a rock magnetic cyclostratigraphy study. Of course, it is important to have a sedimentary section that is likely to have been the result of nearly continuous sedimentation. Small hiatuses and disconformities probably exist in almost any sedimentary sequence, but the proper data analysis, particularly tuning, can help the investigator see through the inevitably incomplete stratigraphic record (Sadler 1981). Platform carbonates and hemipelagic marine sediments have yielded successful rock magnetic

Rock Magnetic Cyclostratigraphy, First Edition. Kenneth P. Kodama and Linda A. Hinnov.
© 2015 John Wiley & Sons, Ltd. Published 2015 by John Wiley & Sons, Ltd.

cyclostratigraphic studies—the Cupido Formation platform carbonates and the Arguis Formation pro-deltaic marls and siltstones covered in Chapter 6 are two good examples. Even continental, fluvial sediments such as the Mississippian Mauch Chunk Formation discussed in Chapter 6 may be able to provide records of short eccentricity and precession, but more rigorous study of similar rocks should be conducted before all terrestrial, fluvial rocks can be trusted to give good cyclostratigraphic results. The discontinuous sedimentation expected in a fluvial environment and the other processes (landsliding, channel avulsion) that could "shred" the climate signal (Jerolmack & Paola 2010) must be considered when conducting a rock magnetic cyclostratigraphic study on fluvial rocks. For this reason, marine sedimentary sequences are preferred because of the greater likelihood of nearly continuous sedimentation.

There should be some kind of independent time control for the stratigraphic section being studied. This is critical for correct identification of astronomically forced climate cycles in the rock magnetic data series. The main power of the cyclostratigraphic technique is its ability to assign time at very high resolution, nominally at the precession scale (~20 kyr), but some time control at a coarser scale is important for a successful study.

Enough time control is needed to identify the longest Milankovitch cycle, usually long eccentricity with a period of ~400 kyr, expected in the data. Some examples of ways that coarse-scale time was assigned to identify astronomically forced cycles in rock magnetic or lithologic cyclostratigraphy studies are magnetostratigraphy (Eocene Arguis Formation), biostratigraphically defined, radiogenically calibrated geologic time scale boundaries (Triassic Daye Formation (Wu et al. 2012), Arguis Formation (Kodama et al. 2010) or the Newark basin depth rank cycles (Olsen & Kent 1996)), or sequence stratigraphy boundaries (Cretaceous Cupido Formation (Hinnov et al. 2013)). Of course, radiogenic ages for ash beds within a stratigraphic section can be an important way of assigning time, but the successful identification of astronomically forced climate cycles will depend on the stratigraphic spacing of the ashes and the errors of the ages. For instance, in the Latemar massif, the ages determined by U-Pb geochronology on zircons extracted from tuff beds had ages that overlapped within their errors, so an estimate of sediment accumulation rate was poorly constrained (Mundil et al. 2003).

Sampling interval and the thickness of section are the next important points to consider when planning a rock magnetic cyclostratigraphy study. The sampling interval is dictated by the Nyquist frequency, i.e., the highest frequency that a cyclostratigraphic study can detect. It must be sampled at least twice per period. If precession is the target for the highest frequency, samples need to be collected at least once per ~10 kyr. This sampling interval is a bare minimum and it is better to sample the precessional cycle at least three or four times or every 5–7 kyr. Given an average sediment accumulation rate of ~10 cm/kyr for rock types that have yielded successful rock magnetic cyclostratigraphies (Table 7.1) would suggest sampling

Table 7.1 Typical average sediment accumulation rates	
Eocene Arguis Fm.	23 cm/kyr
Cretaceous Cupido Fm.	8.2 cm/kyr
Triassic basinal carbonates, Rio Sacuz	5.75 cm/kyr
Triassic Latemar platform carbonates	49 cm/kyr*
Triassic Daye Fm. marine carbonates	3.4 cm/kyr
Stirone R. marine mudstones	24 cm/kyr
Mauch Chunk fluvial red beds	11 cm/kyr
Johnnie Fm. marine clastics/carbonates	4 cm/kyr

*This rate is extrapolated from Rio Sacuz magnetostratigraphy and cyclostratigraphy.

intervals of ~50–70 cm. Aliasing is a problem that must be considered when analyzing a cyclostratigraphy data set.

If cycles with frequencies higher than the assumed precessional frequency are present in the record then the sampling interval chosen to "see" precession will undersample the higher frequencies. The aliasing that will result will lead to spurious low frequencies to emerge in the power spectrum calculated from the data series. One way to identify the effects of aliasing is to sample a limited subsection of the stratigraphic sequence at very small sampling intervals to search for higher frequencies in the record. Wu et al. (2012), for example, sampled the Daye Formation at an amazingly small sampling interval of 2–3 cm and were able to identify sub-Milankovitch scale cycles in the rock magnetic data series, showing that higher than precession frequency cycles are always a possibility. A portable, hand-held susceptibility meter could also be used at a small sampling interval over a subsection of the stratigraphic sequence as a quick check for high frequency cycles.

It is also important to sample enough of the stratigraphic sequence so that the longest cycle targeted for detection, typically the 405 kyr long eccentricity cycle, has room to repeat at least five or six times. For an average sediment accumulation rate of 10 cm/kyr, a 200 m thick section would be ideal. Furthermore, a longer record ensures better bandwidth resolution and narrower spectral peaks in the power spectrum. Certainly, though, smaller stratigraphic sequences can still provide excellent cyclostratigraphic results. The Nopah Range locality of the Neoproterozoic Johnnie Formation was only 40 m thick, but provided a good record of ~8 repetitions of 5 m cycles of short eccentricity and >60 repetitions of 0.6 m cycles of precession. The length of the record, though, will limit the detection of the longest cycle.

The sampling interval for any complementary magnetostratigraphic study is discussed in Chapter 3; however, to repeat here, it is optimal to collect oriented samples from horizons at a stratigraphic spacing that would ensure three horizons per polarity interval. Since the reversal rate of the geomagnetic field varies through geologic time, it is best to know approximately how old the rocks are and to consult the reversal rate of the geomagnetic field at that time period (Merrill et al. 1996; Constable 2000). For the past 170 million years, the maximum reversal rate was about 4–5 reversals per million years, so for a 10 cm/kyr sediment accumulation rate, a magnetostratigraphic sampling interval of 6–7 m would be necessary to acquire three horizons per polarity interval. At each horizon, at least three oriented cores should be collected. If drilling cores in the field isn't a possibility, then oriented blocks can be collected and cores can be drilled from them back at the laboratory. All the techniques needed to drill and collect oriented cores are covered in standard paleomagnetism text books (Butler 1992; Tauxe 2010).

Finally, it is important to determine whether the magnetic mineralogy in the rocks that will likely carry the cyclostratigraphic signal are primary depositional minerals. This is difficult to know without conducting a detailed paleomagnetic study. At a minimum, evidence of the chemical changes associated with present-day weathering, i.e., rusty stains on the rock or yellow brown weathering crusts, should be avoided in the sampling and should be trimmed away during sample preparation in the laboratory.

7.2   Field Sampling

Once the cyclostratigraphy study has been designed, field sampling is fairly straightforward. Compared to a standard paleomagnetic study, many more samples are collected, but the samples do not need to be oriented, so the field procedures are simpler. Usually small hand samples are collected at the stratigraphic interval chosen during the design stage of the study. The hand samples are typically ~5 × 5 × 3 cm blocks that are carefully labeled and usually transported to the laboratory in sample bags, either plastic or cloth. The most important part of the sampling procedure is to keep track of each sample, since the correct sequence of measurements is critical for detecting cycles in the data. Good field notes and prelabeled sample bags can help the field worker avoid missed samples or samples with identical labels collected from different stratigraphic horizons.

Accurately measuring stratigraphic thickness or "measuring section" is a critical part of a rock magnetic cyclostratigraphy study. A commercially available or home-made Jacob's staff is the standard way of measuring section. Once the bedding attitude has been determined with a Brunton compass, the inclinometer on the Jacob's staff is used to orient the staff perpendicular to bedding and the small telescope or sight on the staff is used to determine the location of the next sampling horizon (Figure 7.1). Of course, apparent dip must be calculated if the sampling transect is not perpendicular to the strike of the bedding. Figure 7.2 shows samples collected for a rock magnetic cyclostratigraphy study of the Rainstorm Member of the Neoproterozoic Johnnie Formation. Sampling sites were determined with a Jacob's staff and marked with surveying tape. Samples were broken off with a geologic pick and placed in prelabeled plastic bags. The samples were left

Figure 7.1 Cartoon showing the use of a Jacob's staff to determine stratigraphic intervals for cyclostratigraphy sampling. The operator simply orients the Jacob's staff perpendicular to bedding with a built in inclinometer and determines the location of the next sampling site by looking through the telescope on the Jacob's staff.

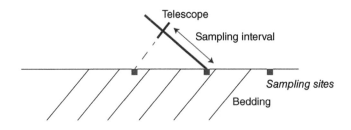

Figure 7.2 Cyclostratigraphy samples collected from the Neoproterozoic Johnnie Formation. Samples were left on the outcrop until the end of the day to be able to check labels against stratigraphic position and for a correct stratigraphic sequence.

on the outcrop until all of a day's samples were collected and the labels were compared against stratigraphic position to double check the label's accuracy before the samples were packed out at the end of the day.

7.3 Laboratory Preparation

After the samples collected in the field have been transported back to the laboratory, they must be prepared for remanence measurements in the rock magnetometer. Rock magnetometers are designed to measure standard size paleomagnetic samples, either plastic sample cubes, usually 7 or 8 cm³ plastic boxes, or 25 mm diameter, 20–23 mm long cores. For cyclostratigraphic studies only the intensity of magnetization is measured, so it is best to average out magnetization heterogeneities as much as possible. For this reason in some studies, the rock samples collected are broken down to millimeter-size pieces and used to fill the plastic sampling boxes that are used for paleomagnetic measurements. Several different boxes are available that range in size from $2 \times 2 \times 2$ cm (8 cm³) to 7 cm³ plastic boxes with rounded edges used by

Figure 7.3 Plastic 2 × 2 × 2 cm sample boxes filled with crushed rock sample from the Cupido Formation.

Figure 7.4 Small 11 mm diameter rock core used for cyclostratigraphic measurements. It is placed in a standard 2 × 2 × 2 cm plastic sample box for remanence measurement in a rock magnetometer.

the IODP (International Ocean Drilling Program), to 2.5 × 2.5 × 1.875 cm boxes used by the older ODP (Ocean Drilling Program). When using these boxes, it is critical that the pieces of rock sample in the boxes can't physically move during measurement, so the sample boxes must be filled completely and the rock pieces need to be forced into the boxes (Figure 7.3). A different approach is to drill small cores (11 mm diameter by ~15 mm long) that will fit in the high field coil of an impulse magnetizer for application of isothermal remanent magnetizations (IRMs). Small plastic cylindrical sample holders (11 mm diameter by 20 mm long) can also be used and filled tightly with small pieces of rock sample. The small diameter coils or plastic cylindrical sample holders must be measured in most rock magnetometers by being placed in standard 2 × 2 × 2 cm plastic sampling boxes (Figure 7.4). Each sample must be weighed, so its remanence measurement can be normalized by its mass. It is easiest to preweigh the sample boxes for the crushed samples before filling the box with rock sample.

7.4 Remanence Measurements

The main point of all the study design, sample collection, and sample preparation is, of course, the measurement of laboratory-applied remanences. These measurements will monitor the variation of magnetic mineral concentration or magnetic mineral particle size throughout the section and these properties will hopefully encode astronomically forced climate change that occurred during the deposition of the sedimentary rocks. The main piece of equipment in a standard paleomagnetism laboratory for remanence measurement is the superconducting rock magnetometer (Figure 7.5), although less expensive, to acquire and maintain, spinner magnetometers can also be used. Remanence measurements must follow a sequence in which anhysteretic remanent magnetization (ARM) is measured before any IRMs are applied to the rock samples. IRMs applied in high fields, $\gtrsim 100$ mT, for example, are difficult to completely demagnetize in a standard

Figure 7.5 2G Enterprises superconducting rock magnetometer used for remanence measurements. Photograph of Lehigh University paleomagnetism laboratory.

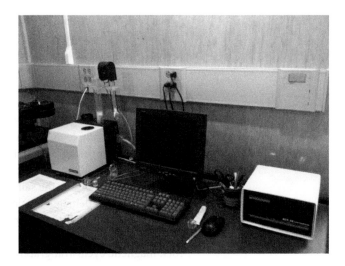

Figure 7.6 KLY-3s Kappabridge susceptibility meter that can be used to measure bulk magnetic susceptibility of cyclostratigraphy samples. Lehigh University paleomagnetism laboratory.

paleomagnetism laboratory, so that any ARM applied after a high-field IRM application will be difficult to measure since the background will be much stronger than the ARM.

The following sequence of measurements is just a suggestion, but it has proven useful in the rock magnetic cyclostratigraphy studies done in the Lehigh University paleomagnetism laboratory:

- Natural remanent magnetization (NRM): NRM measurements won't usually be able to detect climate cycles but they may be able to detect heterogeneously distributed magnetic overprinting either by viscous or chemical effects. Secondary chemical overprints (growth of secondary magnetic minerals) would affect the quality and veracity of the cyclostratigraphic record.

- Magnetic susceptibility (χ): Magnetic susceptibility measurements are typically made on a susceptibility meter, not a rock magnetometer (Figure 7.6). It measures the magnetization induced in a rock sample during the application of small magnetic field, similar in magnitude to the geomagnetic field. A susceptibility measurement is a fast measurement that gives composite concentration variations for all the magnetic minerals: ferromagnetic, paramagnetic, and diamagnetic, in the rock sample. In the Stirone River section study (Gunderson et al. 2012), susceptibility measurements gave the best cyclostratigraphic signal, better than ARM, so susceptibility measurements should always be made in a cyclostratigraphy study. Susceptibility may also be combined with ARM measurements in some studies to monitor magnetic mineral particle size.

- IRM acquisition experiments: IRM acquisition experiments, in which a sample is exposed to higher and higher DC magnetic fields in an impulse magnetizer and its remanence is measured at each field step, can be an important way to identify the ferromagnetic mineralogy in a rock. The IRM acquisition results are usually modeled to determine the coercivity components in a rock (see Chapter 2 and Kruiver et al. (2001)). Usually just a subset of samples distributed throughout the sedimentary section is used for the IRM acquisition measurements. If ARM is used for the main rock magnetic cyclostratigraphy parameter, then different samples must be used for IRM acquisition than the main cyclostratigraphy data set, since an ARM can't be easily measured for a sample that already has a saturation IRM.

- Unblocking temperature–coercivity analysis (Lowrie (1990) test): Another important set of experiments that can help identify the magnetic mineralogy in the samples is the so-called Lowrie test that uses the thermal demagnetization of an IRM for magnetic mineral identification. Magnetic mineral identification is important for determining the rock magnetic parameter that will be used for the rock magnetic cyclostratigraphy. In a Lowrie test, orthogonal IRMs are applied to a sample in different strength magnetic fields. The highest field IRM, typically 1 T, is chosen because it is available in most impulse magnetizers, is applied first, then progressively lower field IRMs are applied in the two perpendicular directions. An intermediate field around 0.3 T may be used, to mark the upper coercivity limit of magnetite, and 0.1 T could be used for the lowest field, since it is the same as the peak alternating field used for ARM application. After application of the orthogonal IRMs, the sample is thermally demagnetized to temperatures that would remove hematite (~680°C), if hematite is suspected in the samples. Similar to the IRM acquisition experiments, the samples subjected to Lowrie tests would not be suitable for a cyclostratigraphic study since the heating could cause chemical changes that would affect the magnetic mineralogy, so a subset of samples is used for the Lowrie test.

- Low-temperature IRM measurements collected with a magnetic property measurement system (MPMS) can also be important data for identifying magnetic minerals from their low-temperature behavior. There are a limited number of MPMS machines available, but there are facilities dedicated to making MPMS machines (and other rock magnetic measurement equipment) available to a wide range of rock magnetic researchers (e.g., Institute of Rock Magnetism at the University of Minnesota). Hysteresis parameters or First Order Reversal Curve (FORC) distributions measured on a vibrating sample magnetometer are also very important ways of identifying the ferromagnetic minerals in a subset of samples from the rock unit being studied.

- Cyclostratigraphy measurements: These remanence measurements are made on a rock magnetometer and are normalized by each sample's mass. The rock magnetic parameter used depends on the magnetic mineralogy identified by the IRM acquisition experiments, the Lowrie tests, MPMS measurements, or hysteresis parameters.

○ If magnetite or magnetite plus iron sulfides are found in the samples, then ARM is the best choice for measuring magnetic mineral concentration variations throughout the section. Usually, a peak field close to 100 mT is the limit for ARM application in most paleomagnetic laboratories and will efficiently activate magnetite and greigite (Fe_3S_4) grains in the sample. The bias field chosen will vary with the equipment used, but is usually a low field close to the geomagnetic field in strength (~50 μT). The results should be normalized by the bias field strength for comparison to ARMs applied with different bias fields. ARMs can be applied by modified alternating field demagnetizers either home-made or commercially available.

○ If hematite is the dominant ferromagnetic mineral in the samples, then an IRM can be used to measure its concentration variations. If both magnetite and hematite are present, first an ARM can be measured followed by application of an IRM. Alternating field demagnetization of the IRM at ~100 mT can remove the effects of magnetite activated by the application of the IRM to isolate the concentration variations of the hematite. An impulse magnetizer which discharges a capacitor through a coil surrounding the sample is used to apply IRMs in fields as high as 1 T for a standard sized paleomagnetic sample ($2 \times 2 \times 2$ cm plastic box) and 5 T for the small 11 mm diameter cores or small plastic cylindrical sample holders.

○ If ancient goethite and hematite are both present in the samples, the goethite to hematite ratio can be measured using Equation (2.5) or Equation (6.1). The sequence of measurements for the G/H ratio is as follows: IRM applied in a high field, usually between 3.5 and 5 T to activate all the ferromagnetic minerals in the sample, thermal demagnetization of the IRM at about 130°C to remove any goethite magnetization, and then alternating field demagnetization at about 100 mT to remove the magnetization of any low coercivity magnetic minerals, like magnetite.

○ Magnetic parameter ratios can be measured to detect slight variations in the amount of magnetite relative to hematite (S-ratio) or the magnetic particle grain size of magnetite (ARM/χ or ARM/SIRM). For magnetic particle size variations, it is important that all the magnetic parameters are carried by the same magnetic mineral.

7.5 Time Series Analysis: Summary of Procedures

Once the rock magnetic measurements have been made and the data series have been constructed, time series analysis is used to detect periodicities in the data. Chapter 4 provides the theoretical background for the analyses used to detect cycles in a data time series; this chapter outlines the practical steps to follow. MATLAB® scripts are included in the Appendix that can help with these analyses; however, two useful software packages will also be referred to here. Analyseries (Paillard et al. 1996) includes routines that evenly subsample a data series, apply a Gaussian filter, calculate Blackman–Tukey, multitaper method (MTM), and maximum entropy spectral estimates, as well as generate theoretical insolation series from Laskar et al. (2004). Analyseries (the current version is 2.0.4.2) is an important tool for time series analysis of rock magnetic data series and is available from the Laboratoire des Sciences du Climat et de l'Environnement (http://www.lsce.ipsl.fr/logiciels/). Almost all of the steps mentioned here can be conducted with Analyseries. The SSA-MTM toolkit (Ghil et al. 2002) is another powerful software package for conducting time series analysis. It offers the calculation of robust red noise (Mann & Lees 1996) for MTM spectral estimates while Analyseries only calculates Thomson (1982) F-tests (with no adaptive weighting) to estimate the significance of spectral peaks.

7.5.1 Preparation of the Data Series

As indicated in Chapter 4, the data series needs to be prepared for time series analysis. The simplest preparation is to resample the series at even intervals because the spectral estimation techniques used all require evenly spaced data. Resampling can be done easily using Analyseries 2.0 provided the caveats mentioned in Chapter 4 are kept in mind. Typically, the average sampling interval used during field sampling sets the resampling interval. Linear interpolation between actual data points has given the best results in our studies.

It is sometimes necessary to remove linear trends or long wavelength variations from the data. Long wavelength trends can yield very low frequency spectral peaks with wavelengths longer than the data series. They may or may not be the reflection of natural phenomena, but they cannot be adequately resolved from the data, so they should be removed before spectral analysis. Analyseries can remove linear trends from the input data series. Taking a moving average of the data series is another way of removing long wavelength trends. MATLAB's *smooth.m* can be used to remove low frequency content; however, care should be taken that a true geologic signal is not removed. Checking the smoothed data series for easily observed periodic signals is important (see Figure 4.4).

Finally, as shown in some of the case studies in Chapter 6, it is sometimes useful to take the $\log_{10}$ of the data series to stabilize the variance of the data and limit the effects of "spikey" data points on the spectral estimate.

7.5.2 Spectral Estimation

Spectral estimation is the centerpiece of this chapter. It is the "time series analysis" that actually produces a power spectrum for the rock magnetic data series. There are three different techniques available in Analyseries:

Blackman–Tukey (B-Tukey), maximum entropy, and MTM, and they all generate power spectra, but in different ways. We have had the most success with MTM analysis, which is described theoretically in Chapter 4. MTM analysis can be conducted by both the Analyseries and SSA-MTM toolkits. The MATLAB script *ftestmtm.m* available in the Appendix can be used for harmonic line detection by MTM analysis while the MATLAB script *pmtm.m* can be used to generate an MTM power spectrum. Chapter 4 shows that the user can select different multitaper taper families by choosing different averaging bandwidths (i.e., 2π, 3π, 4π, 5π, etc.). In most cases, a 2π bandwidth which results in a maximum of three data windows (or a sequence of three Slepian tapers applied to the data series) should yield good results, although changing the bandwidth to be larger or smaller will alter the resolution and confidence in the resulting power spectrum. Different bandwidths should be tested to see the effects on the power spectrum. Using more tapers increases the confidence in the spectral peaks, but causes more smoothing, hence reduces the resolution of the peaks. Therefore, there is a trade-off between resolution (fewer tapers) and confidence (more tapers). The power spectrum should be calculated up to the Nyquist frequency (1/2*sampling interval).

7.5.3   Significance of the Spectral Peaks

One of the important goals of the time series analysis of a rock magnetic data series is to obtain an estimate of the statistical significance of the spectral peaks that emerge in the power spectrum. The question being asked is whether the spectral peak has emerged, statistically, above the background noise in the data series. Analyseries generates harmonic F-tests for MTM spectra that determine the probability that spectral density at any given frequency is associated with true periodic behavior isolated only at that particular frequency compared with the average power of the surrounding frequencies (within the averaging bandwidth). Since F-tests are calculated for all frequencies in the spectrum and are typically plotted on top of the power spectrum, the resulting combined plot can be confusing. It is better to just indicate F-tests for spectral peaks with significant amplitudes. As discussed in Chapter 4, harmonic F-testing can have numerous "false positives" and careful assessment of the power spectrum and F-testing is needed (important guidelines are given in Thomson (2009)).

A more general approach for determining spectral peak significance is the generation of a red noise spectrum for the data series and different confidence levels above that red noise. The SSA-MTM toolkit will generate a robust red noise (Mann & Lees 1996) spectrum with 90, 95, and 99% confidence intervals as part of its MTM spectral estimation. Spectral peaks that rise, for instance, above the 95% confidence interval for the red noise are statistically distinct from the background red noise model with 95% confidence. The red noise model assumes that the noise in the data has a memory by one time step (AR(1) process: first-order autoregressive process), i.e., it is a way of modeling the natural inertia, or hysteresis, in a system and is reasonable for geological and climate processes. The MATLAB script *rednoise.m* in the Appendix can also be used to generate an AR(1) red noise spectrum for identification of significant spectral peaks in the MTM power spectrum.

7.5.4   Evolutionary Spectrogram

All stratigraphic data series suffer from uneven sediment accumulation rates and hiatuses in the record. Furthermore, rock magnetic cyclostratigraphic series may suffer from uneven quality of the recording of natural periodic behavior, i.e., global climate cycles, due, for example, to postdepositional changes in the magnetic mineralogy or variations in the supply or transport of magnetic minerals to the depositional basin. One good way to investigate whether these processes have affected the rock magnetic cyclostratigraphic data series is to calculate an evolutionary spectrogram (Figure 4.24) using MATLAB script *evofft.m* (Appendix). (Analyseries will also calculate evolutionary spectrograms.) The script calculates the periodogram of a portion (window) of the data series and steps the window through the series, plotting the resulting spectra as a function of the step increment. The user selects the data window size, which can be no larger than the total data series but should be much smaller, and the increments at which the window is stepped through the data series that are much smaller than the data window. The rock magnetic data evolutionary spectrogram can be compared to an evolutionary spectrogram generated from the theoretical insolation series of similar age (Laskar et al. 2004) to see if changes in the amplitudes of Milankovitch spectral peaks match (Figure 4.24) thus providing evidence to support identification of astronomically forced climate cycles.

7.5.5   Tuning and Filtering

As indicated in Chapter 4, filtering can be an important tool for examining the frequency content of the rock magnetic cyclostratigraphic data series. For instance, filtering the data series at a suspected precessional frequency and checking its amplitude modulation for eccentricity frequencies is a useful way to identify Milankovitch cycles in the data (see the Cupido Fm. and Daye Fm. studies in Chapter 6). Furthermore, filtering a data set at a given Milankovitch frequency can facilitate tuning to the theoretical insolation variations for that Milankovitch frequency at the age of the sedimentary sequence. The tuning can help remove the effects of slight changes in sediment accumulation rate and small hiatuses in the record, with the potential of enhancing the power spectrum. Analyseries can apply Gaussian band pass filters to a data series; *gaussfilter.m* and *tanerfilter.m* (Appendix) provide stable alternatives.

Tuning of a data series means simply that the peaks or troughs for a given cycle in the data are adjusted in their spacing to match either the theoretical insolation for that cycle (precession, obliquity, or eccentricity) calculated from Laskar et al. (2004, 2011) or regularly spaced intervals set to a cycle expected to be regular throughout geologic time (the 405 kyr long eccentricity cycle, for example). Analyseries can generate theoretical insolation series based on Laskar et al. (2004). The Appendix offers a way of calculating theoretical precession and obliquity insolation variations based on Laskar et al. (2011). Analyseries's Linage function can be used to tune a filtered data series (distorted series) to the theoretical insolation series (reference series). A pointer series can also be generated to adjust the unfiltered data series to its new "tuned" spacing using Analyseries's Age Scale function. In MATLAB, *picktune.m* (Appendix) has now been developed as an alternative to the Linage function.

7.6 Identifying Astronomically Forced Climate Cycles

Of course, the whole point of a rock magnetic cyclostratigraphy study is the identification of astronomically forced climate cycles in the rock magnetic record. These cycles have well-defined durations and when properly identified will provide a continuous, high-resolution chronostratigraphy throughout the sedimentary sequence. The most desirable way to acquire a high-resolution chronostratigraphy is to be able to tune the rock magnetic cyclostratigraphy to theoretical variations in insolation for eccentricity, obliquity, and precession. Laskar et al. (2011) give the most recent theoretical calculations for astronomically forced insolation variations. They indicate that the solutions are stable back to ~50 Ma. For times before that, Laskar et al. (2011) indicate that the 405 kyr, long eccentricity cycle has been stable, at least back to 250 Ma, but probably even further back in geologic time.

Therefore, the best way to identify astronomically forced climate cycles is to use absolute time control at a coarse scale to be able to identify a long-period astronomically forced cycle, usually the 405 kyr long eccentricity cycle. Once that cycle is observed, the rock magnetic cyclostratigraphy data series is tied to time, by tuning it to that period. If the sedimentary rocks are younger than 50 Ma, the data series can be band pass filtered at the possible 405 kyr period and the filtered record can be tied to theoretical insolation at the 405 kyr period. For time periods before 50 Ma, the 405 kyr filtered record can simply be tuned by evenly spacing the possible long eccentricity periods at 405 kyr intervals. After the rock magnetic cyclostratigraphy record is tied to time in one of these manners, time series analysis of the unfiltered data series will determine whether statistically significant spectral peaks emerge in the power spectrum. This approach was used successfully for the Eocene Arguis Formation (Kodama et al. 2010). Short eccentricity and precession, for example, might emerge if the sediments were deposited at low latitudes, where precession-scale insolation variations are expected to drive the monsoon. For icehouse periods of Earth's history, obliquity might be observed due to variations in sea level driven by changes in global ice volume because insolation variations at the obliquity timescale are important at high latitudes. If higher frequency astronomically -forced cycles emerge in the power spectrum after tuning to 405 kyr long eccentricity, then the long eccentricity cycle was correctly identified and a high resolution chronostratigraphy can be generated.

Assigning time at a coarse scale can be accomplished by several methods. One powerful way, covered in this book, is to determine a magnetostratigraphy for the sedimentary sequence. At some points in Earth history, the reversal rate of the geomagnetic field was as high as 4–5 reversals per million years, already giving nearly 200 kyr time resolution. At other times, the reversal rate diminished to as low as no reversals per million years in the Cretaceous (126–84 Ma), when magnetostratigraphy is not possible. The geomagnetic field also remained one polarity (reversed) for tens of millions of years during the Late Paleozoic Kiaman superchron, so it is important to know the age of the sedimentary sequence being studied to determine if magnetostratigraphy is even possible. When magnetostratigraphy is an option, it is an excellent way of assigning coarse time to the sedimentary sequence to identify orbitally forced cycles. Magnetostratigraphy was used in this way for the Eocene Arguis Formation (Kodama et al. 2010) and the Stirone River sediments (Gunderson et al. 2012).

Biostratigraphy can also be used to assign coarse resolution time to the sedimentary record under study. Wu et al. (2012) used the conodont biostratigraphy of the rocks, which had been correlated globally and calibrated radiogenically, in this manner to identify the 405 kyr long eccentricity cycles. They also used their coarse age model to identify higher frequency astronomically forced cycles and tuned them to short eccentricity and to precession.

Olsen and Kent (1996) identified the ~400 kyr long eccentricity cycle in the lithologic cyclostratigraphy of the Newark Basin lacustrine rocks by assigning time at three different geologic time-scale boundaries. According to Olsen and Kent (1996), there are three biostratigraphically defined boundaries in the Newark Basin rocks: the Triassic–Jurassic boundary, the Norian–Rhaetian boundary, and the Carnian–Norian boundary, mainly identified by pollen and spore assemblages. Olsen and Kent (1996) recognize long and short lithologic cycles that they interpret to indicate variations in the depth of the lake that deposited the sediments. The longer McLaughlin cycle, which modulates the shorter Van Houten cycles, is shown to have a duration of nearly 400 kyr (397.7 ± 58.5 kyr) based on the ages for the geologic time-scale boundaries. In this manner, Olsen and Kent (1996) very nicely demonstrated that the McLaughlin cycles were driven by long eccentricity and could calibrate their lithologic cyclostratigraphy accordingly. A similar approach could be used to identify long eccentricity in a rock magnetic cyclostratigraphy record.

Another approach is to use the fourth order sequence boundaries that mark the change from transgression to regression in sequence stratigraphy.

Long eccentricity was assumed to drive the fourth order sequence stratigraphy cycles in the Cretaceous Cupido Formation platform carbonates (Figure 6.6, (Hinnov et al. 2013)). This assumption was supported by the emergence of spectral peaks with 100, 44, and 20 kyr periods, once the rock magnetic cyclostratigraphy data series was tied to time assuming the sequence boundaries were spaced at 405 kyr intervals.

Of course, radiometric ages from volcanic ash fall beds throughout a sedimentary sequence could easily give enough time control to estimate the sediment accumulation rate and hence the approximate duration of cycles recognized in the time series analysis of the rock magnetic data series. The distribution of the ash fall beds in the stratigraphic section and the errors on the ages will limit the accuracy and resolution of the age model. For instance, in the Latemar Triassic carbonate platform sequence of the Dolomites in northern Italy, three volcanic ash fall beds are spaced over ~250 m in >600 m thick section. The ages range between 241.2 and 242.6 Ma with overlapping error limits. A range of sediment accumulation rates calculated from the ages varies by a factor of two from 14 to 28 cm/kyr (Mundil et al. 2003). Mundil et al. (2003) do present a better-constrained sediment accumulation rate, but by bringing in additional geochronologic data by correlation to a different sedimentary section.

When there is no good control on absolute time available, the ratio of the periods for the cycles that emerge in the spectral analysis can be used as evidence for astronomically forced climate cycles. Observations of 20:5:2:1 ratios, which are the approximate ratios for long eccentricity:short eccentricity:obliquity:precession (400:100:40:20), or some portion of them, can suggest Milankovitch forcing. This approach is the least desirable way to detect astronomically forced cycles since misidentification can and will happen. The 5:1 bundling of the Latemar facies cyclicity, for instance, was the primary evidence for short eccentricity:precession driving the sedimentary cyclicity until geochronology and magnetostratigraphy suggested otherwise.

References

Butler, R.F. (1992) *Paleomagnetism: Magnetic Domains to Geologic Terranes*, 319 pp. Blackwell Scientific Publications, Boston.

Constable, C. (2000) On rates of occurrence of geomagnetic reversals. *Physics of the Earth and Planetary Interiors*, *188*, 181–193. DOI:10.1016/s0031-9201(99)00139-9.

Ghil, M., Allen, M.D., Dettinger, K.I., Kondrashov, D., Mann, M., Roberts, A.P., Saunders, A., Tian, Y., Varadi, F., & Yiou, P. (2002) Advanced spectral methods for climatic time series. *Reviews of Geophysics*, *40*, 3.1–3.41. DOI:10.1029/2000RG000092.

Gunderson, K.L., Kodama, K.P., Anastasio, D.J., & Pazzaglia, F.J. (2012) Rock-magnetic cyclostratigraphy for the Late Pliocene-Early Pleistocene Stirone section, Northern Apennnine mountain front, Italy. *Geological Society, London, Special Publications*, *373*, 26. DOI:10.1144/SP373.8.

Hinnov, L.A., Kodama, K.P., Anastasio, D.J., Elrick, M., & Latta, D.K. (2013) Global Milankovitch cycles recorded in rock magnetism of the shallow marine lower

Cretaceous Cupido Formation, northeastern Mexico. In: Jovane, L., Herrero-Bervera, E., Hinnov, L.A., and Housen, B.A. (eds), *Magnetic Methods and the Timing of Geological Processes*, p. 15. Geological Society, London.

Jerolmack, D.J. & Paola, C. (2010) Shredding of environmental signals by sediment transport. *Geophysical Research Letters*, 37. DOI:10.1029/2010GL044638.

Kodama, K.P., Anastasio, D.J., Newton, M.L., Pares, J., & Hinnov, L.A. (2010) High-resolution rock magnetic cyclostratigraphy in an Eocene flysch, Spanish Pyrenees. *Geochemistry, Geophysics, Geosystems, 11.* DOI:10.1029/2010GC003069.

Kruiver, P.P., Dekkers, M.J., & Heslop, D. (2001) Quantification of magnetic coercivity components by the analysis of acquisition curves of isothermal remanent magnetisation. *Earth and Planetary Science Letters, 189,* 269–276. DOI:10.1016/s0012-821x(01)00367-3.

Laskar, J., Robutel, P., Joutel, F., Gastineau, M., Correia, A.C.M., & Levrard, B. (2004) A long term numerical solution for the insolation quantitties of the Earth. *Astronomy and Astrophysics, 428,* 261–285. DOI:10.1051/0004-6361:20041335.

Laskar, J., Fienga, A., Gastineau, M., & Manche, H. (2011) La2010: A new orbital solution for the long-term motion of the Earth. *Astronomy and Astrophysics, 532 (A89).* DOI:10.1051/0004-6361/201116836.

Lowrie, W. (1990) Identification of ferromagnetic minerals in a rock by coercivity and unblocking temperature properties. *Geophysical Research Letters, 17,* 159–162. DOI:10.1029/gl017i002p00159.

Mann, M. & Lees, J. (1996) Robust estimation of background noise and signal detection in climatic time series. *Climate Change, 33,* 409–445. DOI:10.1007/bf00142586.

Merrill, R.T., McElhinny, M.W., & McFadden, P.L. (1996) *The Magnetic Field of the Earth, Paleomagnetism, the Core, and the Deep Mantle.* Academic Press, San Diego.

Mundil, R., Zuhlke, R., Bechstadt, T., Peterhansel, A., Egenhoff, S.O., Oberli, F., Meirer, M., Brack, P., & Rieber, H. (2003) Cyclicities in Triassic platform carbonates: Synchronizing radio-isotopic and orbital clocks. *Terra Nova, 15,* 81–87. DOI:10.1046/j.1365-3121.2003.00475.x.

Olsen, P.E. & Kent, D.V. (1996) Milankovitch climate forcing in the tropics of Pangea during the Late Triassic. *Palaeogeography, Palaeoclimatology, Palaeoecology, 122,* 1–26. DOI:10.1016/0031-0182(95)00171-9.

Paillard, D., Labeyrie, L., & Yiou, P. (1996) Macintosh program performs time-series analysis. *Eos, Transactions, American Geophysical Union, 77,* 379. DOI:10.1029/96eo00259.

Sadler, P.M. (1981) Sedimentation rates and the completeness of stratigraphic sections. *Journal of Geology, 89,* 569–584. DOI:10.1086/628623.

Tauxe, L. (2010) *Essentials of Paleomagnetism,* 489 pp. University of California Press, Berkeley.

Thomson, D.J. (1982) Spectrum estimation and harmonic analysis. *IEEE Proceedings, 70,* 1055–1096. DOI:10.1109/proc.1982.12433.

Thomson, D.J. (2009) Time-series analysis of paleoclimate data. In: Gornitz, V. (ed), *Encyclopedia of Paleoclimatology and Ancient Environments,* pp. 949–959. Springer, New York.

Wu, H., Zhang, S., Feng, Q., Jiang, G., Li, H., & Yang, T. (2012) Milankovitch and sub-Milankovitch cycles of the early Triassic Daye Formation, South China and their geochronological and paleoclimatic implications. *Gondwana Research, 22,* 748–759. DOI:10.1016/j.gr.2011.12.003.

Appendix

Abstract: This appendix contains detailed information about the applications and operations carried out in this book. Section A.1 lists the MATLAB® functions used in this book; Section A.2 lists and describes the MATLAB scripts used in this book; Section A.3 provides specific command strings that were used to create selected figures; Section A.4 explains how to obtain, compile, and use FORTRAN code to calculate customized precession and obliquity with Laskar astronomical models; Section A.5 is a list of key resources; Section A.6 lists cited references.

A.1 MATLAB Functions

Functions from the MATLAB, Curve Fitting, and Signal toolboxes used in this book (usage demonstrated in Section A.3):

`angle.m`—calculates phase from a complex variable

`bartlett.m`—calculates the Bartlett taper (window)

`cat.m`—concatenates two arrays

`chi2pdf.m`—calculates the χ^2 probability distribution function

`diff.m`—takes the first difference of a vector

`dpss.m`—calculates discrete prolate spheroidal sequences (prolate multitapers)

`fcdf.m`—calculates the F-distribution cumulative distribution function

`fft.m`—fast Fourier transform, uses

`hann.m`—calculates the Hann taper (window)

`interp1.m`—performs interpolation on unequally spaced series

`padarray.m`—pads an array with zeros

`pmtm.m`—computes the Thomson prolate multitaper power spectrum

`randn`—computes random numbers with normal (Gaussian) probability distribution

`random`—computes random numbers with customized probability distributions

`rectwin`—calculates the rectangular (boxcar) taper (window)

`smooth.m`—smooths a series with multiple options

Rock Magnetic Cyclostratigraphy, First Edition. Kenneth P. Kodama and Linda A. Hinnov.
© 2015 John Wiley & Sons, Ltd. Published 2015 by John Wiley & Sons, Ltd.

spectrum.mtm.m—computes the Thomson multitaper power spectrum (uses pmtm.m)

spectrum.periodogram.m—periodogram spectral estimation with taper (window) options

spectrum.welch.m—computes the Welch spectral estimator, including WOSA

wvtool.m—the Window Visualization Tool, a (GUI) to analyze windows (tapers).

xcorr.m—computes the cross-correlation function

A.2 MATLAB Scripts by Authors and Colleagues

Scripts used in this book with download information (usage demonstrated in Section A.3):

alphanoise.m—theoretical power law noise spectra as a function of a, Linda Hinnov (Johns Hopkins University):
https://jshare.johnshopkins.edu/lhinnov1/scripts/

arnoisemodel.m—theoretical order 1 autoregressive noise spectra as a function of r, Linda Hinnov (Johns Hopkins University):
https://jshare.johnshopkins.edu/lhinnov1/scripts/

chisquare.m (uses chi2pdf.m)—calculates χ^2_n distributions, Linda Hinnov (Johns Hopkins University):
https://jshare.johnshopkins.edu/lhinnov1/scripts/

cmtm.m, (and cohbias.m, cohbias.mat) Prolate multitaper spectral coherency and cross-phase analysis, Peter Huybers (Harvard University):
http://www.people.fas.harvard.edu/~phuybers/Mfiles/index.html

depthtotime.m—converts a stratigraphic series to a time series with an age model, Linda Hinnov (Johns Hopkins University):
https://jshare.johnshopkins.edu/lhinnov1/scripts/

estdof.m—estimates effective degrees of freedom of a periodogram, Linda Hinnov (Johns Hopkins University):
https://jshare.johnshopkins.edu/lhinnov1/scripts/

evofft.m—evolutionary FFT modulus, Florian Maurer (University of Amsterdam) and Linda Hinnov (Johns Hopkins University):
https://jshare.johnshopkins.edu/lhinnov1/scripts/

ftestmtm.m—prolate multitaper harmonic line estimation with adaptive-weights and F-test, Jeffrey Park (Yale University) and Linda Hinnov (Johns Hopkins University):
https://jshare.johnshopkins.edu/lhinnov1/scripts/

`gaussfilter.m`—Gauss filter, Linda Hinnov (Johns Hopkins University) and Karthikeyan Ramamurthy (Arizona State University):
https://jshare.johnshopkins.edu/lhinnov1/scripts/

`hilbertsignal.m`—quadrature analysis of filtered signals, Florian Maurer (University of Amsterdam) and Linda Hinnov (Johns Hopkins University):
https://jshare.johnshopkins.edu/lhinnov1/scripts/

`markovseries.m`—red noise series, Linda Hinnov (Johns Hopkins University):
https://jshare.johnshopkins.edu/lhinnov1/scripts/

`mtmcoherency.m`—prolate multitaper squared-coherency and cross-phase spectrum estimation with adaptive weighting, Linda Hinnov (Johns Hopkins University):
https://jshare.johnshopkins.edu/lhinnov1/scripts/

`mtmcoherencystats.m`—theoretical and jackknifed statistics for prolate multitaper squared-coherency and cross-phase spectrum estimation, Linda Hinnov (Johns Hopkins University):
https://jshare.johnshopkins.edu/lhinnov1/scripts/

`mtmdofs.m`—effective degrees of freedom for adaptive-weighted prolate multitaper spectral estimates, Linda Hinnov (Johns Hopkins University):
https://jshare.johnshopkins.edu/lhinnov1/scripts/

`picktune.m`—picks specific points in a time series for tuning; store as age model, Linda Hinnov (Johns Hopkins University):
https://jshare.johnshopkins.edu/lhinnov1/scripts/

`Rednoise.m` (and `rhoAR1.m`, `redconf.m`, `conflevel.m`, `chi2invPMTK.m`)—classical red noise with Monte Carlo modeling, Dorothée Husson (Northwestern University):
http://www.mathworks.com/matlabcentral/fileexchange/45539-rednoiseconfidencelevels

`staircase.m`—interpolates facies rank series, Linda Hinnov (Johns Hopkins University):
https://jshare.johnshopkins.edu/lhinnov1/scripts/

`tanerfilter.m`—Taner filter, Linda Hinnov (Johns Hopkins University):
https://jshare.johnshopkins.edu/lhinnov1/scripts/

`daily_insolation.m`, insolation equation calculations, Peter Huybers (Harvard University) and Ian Eisenman (University of San Diego):
http://www.ncdc.noaa.gov/paleo/pubs/huybers2006b/huybers2006b.html
http://eisenman.ucsd.edu/code/

A.3   Command Strings used for Selected Figures

The following command strings reproduce selected figures in the text. The corresponding plots displayed in Chapters 4 and 5 have been edited for clarity and aesthetics. The Eocene Arguis Formation datasets that are analyzed, "arguisgrain.txt" and arguisarm.txt from Kodama et al. (2010) may be downloaded—as well as command strings for the remaining figures—from:

https://jshare.johnshopkins.edu/lhinnov1/scripts/

Figure 4.2 Staircase interpolation of Arguis grain size stratigraphic series: Input Arguis Formation grain size series in a two-column array arguisgrain, with stratigraphic height in column 1 and grain size (integer from 3 = fine to 9 = coarse) in column 2.

```
dint=0.5;
[stratint,grainint]=staircase(arguisgrain,dint);
figure;plot(stratint,grainint);
```

Figure 4.3a Assessing the variable sample rate of ARM stratigraphic series: Input Arguis Formation ARM series is in the two-column array arguisarm, with stratigraphic height in column 1 and anhysteretic remanent magnetism (ARM) in column 2. These are separated into individual vectors, strat and arm, and then the first difference of strat is taken to assess the sample rate.

```
strat=arguisarm(:,1);
arm=arguisarm(:,2);
len=length(strat);
dstrat=diff(strat);
strat1=strat(1):strat(len-1);
figure;plot(strat1,dstrat);
```

Figure 4.3b and c Linear interpolation of the ARM stratigraphic series to the mean sample rate (0.6757 m): Mean sample rate is assessed; series is interpolated using interp1.m to the mean sample rate; original and interpolated ARM stratigraphic series are plotted together.

```
dmean=mean(diff(strat));
dintmean=2.8:dmean:795.4;
armdintmean=interp1(strat,arm,dintmean);
figure;plot(strat,arm);  hold  all;  plot(dintmean,
armdintmean);
```

Figure 4.3d Spline interpolation of the ARM stratigraphic series to a high-resolution sample rate (0.05 m): A high-resolution sample rate is set at $\Delta d = 0.05$; ARM series is interpolated using interp1.m with the "spline" option; the original and interpolated ARM stratigraphic series are plotted together.

```
dhigh=0.05;
dinthigh=2.8:dhigh:795.4;
```

```
armdinthighspline=interp1(strat,arm,dinthigh,&rsqu
o;spline&rsquo;);
figure;plot(strat,arm); hold all; plot(dinthigh,ar
mdinthighspline);
```

Figure 4.4 Smoothing options for the $\Delta d = 0.05$ m linearly interpolated ARM stratigraphic series: A $\Delta d = 0.05$ m linearly interpolated Arguis ARM series is smoothed in three different ways over an 80 m window using smooth.m and plotted together.

```
armdinthigh=interp1(strat,arm,dinthigh);
span=80/0.05;
armmoving=smooth(armdinthigh,span,'moving');
armloess=smooth(armdinthigh,span,'loess');
armrloess=smooth(armdinthigh,span,'rloess');
figure; plot(dinthigh,arminthigh,'red');
hold all; plot(dinthigh,armrloess,'grey');
hold all; plot(dinthigh,armmoving,'green');
hold all; plot(dinthigh,datalowess,'blue');
hold all; plot(dinthigh,dataloess,'black');
```

Figure 4.6 Gauss and Taner filter bode plots. Create an impulse function of length 4096 using padarray.m, set dt = 1 and passband to fl = 0.035, fc = 0.050, fh = 0.065, pass through gaussfilter.m and tanerfilter.m, then apply fft.m and calculate modulus and phase plots.

```
impulse=padarray(1,4095,'post');
fc=0.05;fl=fc-0.015;fh=fc+0.015;dt=1.;
gaussbandx=gaussfilter(impulse,dt,fc,fl,fh);
tanerbandx=tanerfilter(impulse,dt,fc,fl,fh);
tanerfft=fft(complex(tanerbandx));
tanermag=abs(tanerfft.^2);
tanerfaz=angle(tanerfft);
f=0.:1/4096:1-1/4096;
figure;plot(f,tanermag);figure;plot(f,tanerfaz);
gaussfft=fft(complex(gaussbandx));
gaussmag=abs(gaussfft.^2);
gaussfaz=angle(gaussfft);
figure;plot(f,gaussmag);figure;plot(f,gaussfaz);
```

Figure 4.7 Application of Gauss and Taner filters to extract precession index from insolation: In *Analyseries 2.0.4.2*, with the La2004 model, create 65° North mean summer insolation from March 21 to September 21, 36,000–40,000 ka with a 1 kyr step; select and save data. Create and save precession index for the same time interval using the same procedure.

```
% Import insolation file as 'data' (use File menu)
time=data(:,1);
summer=data(:,2);
figure;plot(time,summer); % Figure 4.7a
```

```
r=randn(size(summer));
sdsummer=std(summer);
sdr=std(r);
noisysummer=summer+sdsummer*r/sdr;
figure;plot(time,noisysummer); % Figure 4.7b
fc=0.05;fl=fc-0.015;fh=fc+0.015;dt=1.;
gaussnoisysummer=gaussfilter(noisysummer,dt,fc,fl,fh);
figure;plot(time,gaussnoisysummer); % Figure 4.7c
tanernoisysummer=tanerfilter(noisysummer,dt,fc,fl,fh);
figure;plot(time,tanernoisysummer); % Figure 4.7d
% Import precession index file as 'data' (use File menu)
time=data(:,1);
precession=data(:,2);
figure;plot(time,precession); % Figure 4.7e
```

Note: Windows users can compute insolation and precession index series with the online insolation calculator: http://www.imcce.fr/Equipes/ASD/insola/earth/online/

Figure 4.8 Example of a real-valued time series of length 512 with two closely spaced frequencies 0.050 and 0.055 and its FFT; plot the modulus of the FFT:

```
t=1:1:512;
signal=sin(2*pi*t*0.05)+sin(2*pi*t*0.055);
figure;plot(signal);
y=fft(complex(signal));
mag=2*abs(y);
figure;plot(mag);
figure; plot(mag,'.-','markeredgecolor','k','marker
facecolor','k');
```

Figure 4.10 Effects of the Dirichlet (no taper), Bartlett, and Hann tapers on the FFT are demonstrated.

```
tt=1:1:2048;
nn=length(tt);
signal=sin(2*pi*tt*0.05)+sin(2*pi*tt*0.055);
figure;plot(tt,signal);
barttaper=bartlett(nn);
bartsignal=barttaper.*signal'; % note transpose (')
of signal
figure;plot(tt,bartsignal);
hanntaper=hann(nn);
hannsignal=hanntaper.*signal';
figure;plot(tt,hannsignal);
% Dirichlet
xx=complex(signal);
y=fft(xx);
mag=2*abs(y);
```

```
% Bartlett
xx=complex(bartsignal);
y=fft(xx);
bartmag=2*abs(y);
bartmagcorrect=1.5*bartmag; % leakage correction
% Hann
xx=complex(hannsignal);
y=fft(xx);
hannmag=2*abs(y);
hannmagcorrect=1.6*hannmag; % leakage correction
% Create frequency scale and plot the three moduli
f1=0:1/nn:1/2;
f2=1/2-1/nn:-1/nn:1/nn;
f=cat(1,f1',-1*f2');
% Note: MATLAB plot.m arranges frequency scale from
min to max
figure;plot(f,mag);
hold all;plot(f,bartmagcorrect);
hold all;plot(f,hannmagcorrect);
```

Figure 4.11 Plotting the χ^2_n distribution for n = 1, 2, …, 20:
```
x = 0:0.2:100;
ychi=chisquare(x);
figure;plot(x,ychi);
```

Figure 4.12 Hann-tapered periodograms with no averaging and different averaging with the Welch overlapping segment averaging (WOSA) technique:
```
%compute unaveraged periodograms
h = spectrum.periodogram('rectangular');
hopts = psdopts(h,signal);
set(hopts,'SpectrumType','twosided');
hpsdirich = psd(h,signal,'ConfLevel',0.95);
figure;plot(hpsdirich);
h = spectrum.periodogram('hann');
hpsdhann = psd(h,signal,'ConfLevel',0.95);
figure;plot(hpsdhann);
%compute WOSA periodograms
h = spectrum.welch('hann',1024);
hpsdwelchhann1024 = psd(h,signal,'ConfLevel',0.95);
figure;plot(hpsdwelchhann1024);
h = spectrum.welch('hann',512);
hpsdwelchhann512 = psd(h,signal,'ConfLevel',0.95);
figure;plot(hpsdwelchhann512);
```

Figure 4.13 Application of zero-padding:
```
tt=1:1:2048;
```

```
nn=length(tt);
signal=sin(2*pi*tt*0.05)+sin(2*pi*tt*0.055);
y=fft(complex(signal));
mag=2*abs(y);
f1=0:1/nn:1/2;
f2=1/2-1/nn:-1/nn:1/nn;
f=cat(1,f1',-1*f2');
figure;plot(f,mag); hold all;
%beware of transpose character; use MATLAB command
line version (');
signal=signal';
signalpad=padarray(signal,10*2048-2048,'post');
nnpad=length(signalpad);
ypad=fft(complex(signalpad));
magpad=2*abs(ypad);
f1=0:1/nnpad:1/2;
f2=1/2-1/nnpad:-1/nnpad:1/nnpad;
fpad=cat(1,f1',-1*f2');
plot(fpad,magpad);
```

Figure 4.14 Computation of tapered Blackman–Tukey correlograms:

```
%calculate  autocorrelation  function  of  signal
(see Fig. 4.10) and plot
rhos=xcorr(signal);
nn=length(rhos);
rhos=rhos';
trhos=-2047:1.:2047;
figure;plot(trhos,rhos/max(rhos));
%FFT of autocorrelation function; plot (Dirichlet)
modulus
xx=complex(rhos);
y=fft(xx);
magrhos2=2*abs(y);
f1=0:1/nn:1/2;
f2=1/2-1/nn:-1/nn:0;
frhos=cat(1,f1',-1*f2');
%correct normalization may be just nn
figure; plot(frhos,magrhos2/(nn/2));
%apply Hann taper M=N to autocorrelation function
and plot
hannlag=hann(nn);
hannrhos=hannlag.*rhos;
figure;plot(trhos,hannrhos/max(hannrhos));
%FFT of Hanned autocorrelation function; plot M=N
Hanned modulus
xx=complex(hannrhos);
```

```
y=fft(xx);
maghannrhos2=2*abs(y);
figure;plot(frhos,maghannrhos2/(nn/2));
%apply Hann taper M=0.5N to autocorrelation function
and plot
hannlag50=hann(round(nn/2));
%center the 50% Hann taper in hannwindow
hannwindow=[ ];
hannwindow=zeros(1,nn);
hannwindow=hannwindow';
nstart=nn/4;
nend=3*nn/4;
m=1;
for n=1:nn
if n >= nstart & n < nend
hannwindow(n)=hannlag50(m);
m=m+1;
end
end
%apply the 50% Hann taper stored in hannwindow
hannrhos50=hannwindow.*rhos;
figure;plot(trhos,hannrhos50/max(hannrhos50));
%FFT of 50% Hanned autocorrelation function; plot
modulus
xx=complex(hannrhos50);
y=fft(xx);
maghannrhos502=2*abs(y);
figure;plot(frhos,maghannrhos502/(nn/2));
```

Figure 4.17a Signal with white noise and red noise. Note: generated noise series will not be identical to those in the figure.

```
tt=1:1:2048;
signal=sin(2*pi*tt*0.05)+sin(2*pi*tt*0.055);
noise5=5*randn(2048,1);
rho=0.9;
markov5=markovseries(rho,noise5);
% transpose so values are in a column
markov5=markov5';
signal=signal';
signalnoise5=signal+noise5;
signalmarkov5=signal+markov5;
```

Figure 4.18 Theoretical autoregressive (AR) and $1/f^{\alpha}$ spectral noise models:

```
%arnoisemodel.m for theoretical AR noise models with
rho from 0.0 to 0.9
[farnoise,arnoise]=arnoisemodel();
figure;plot(farnoise,arnoise);
```

```
%alphanoise.m for 1/f^alpha noise models, alpha = 1
and 2
% Note: outputs alpha1 and alpha2 are time series
realizations;
%         mag1 and mag2 are the theoretical noise
spectra
npts=2048;
fmag=0:1/npts:0.5-1/npts;
fmag=fmag';
[alpha1,mag1]=alphanoise(1,npts);
[alpha2,mag2]=alphanoise(2,npts);
mag1=mag1';
mag2=mag2';
mag12=mag1.^2;
mag22=mag2.^2;
mag12a=(length(mag12)/sum(mag12)) * mag12;
mag22a=(length(mag22)/sum(mag22)) * mag22;
figure;plot(fmag,mag12a);holdall;plot(fmag,mag22a)
```

Figure 4.19 Hypothesis testing of the signal+red noise time series in **Figure 4.17**.

```
dt=1;num=1000;nw=4;
[Mspecred,specred,po,fd,fr,tabMC,tabchi,tabtchi,th
eored,rho]=redconf(signalmarkov5,tt,dt,num,nw);
figure;plot(fd,po);hold all; plot(fd,theored);hold
all;plot(fd,tabtchi);
```

A.4 Computation of the Obliquity and Precession Index

For the La2010a (or b, c, or d) solution calculations for the obliquity and precession index can be made using FORTRAN programs as follows:

(1) The Laskar et al. (1993) FORTRAN code and input parameters files may be downloaded from the online VizieR Catalogue Service:

http://vizier.cfa.harvard.edu/ftp/cats/VI/63/prepa.f
http://vizier.cfa.harvard.edu/ftp/cats/VI/63/prepa.par
http://vizier.cfa.harvard.edu/ftp/cats/VI/63/integ.f
http://vizier.cfa.harvard.edu/ftp/cats/VI/63/integsub.f
http://vizier.cfa.harvard.edu/ftp/cats/VI/63/integ.par
http://vizier.cfa.harvard.edu/ftp/cats/VI/63/climavar.f
http://vizier.cfa.harvard.edu/ftp/cats/VI/63/climavar.par

Mac users must apply mac2unix (*one time only*) to convert all files after downloading, which may be obtained here:
http://sourceforge.net/projects/cs-cmdtools/files/

(2) Concatenate `integ.f` and `integsub.f` into a single file `integall.f`.

(3) Compile `prepa.f`, `integall.f` and `climavar.f` using `gfortran` (which must be preinstalled, see http://cran.r-project.org/bin/macosx/tools/ to acquire free `gfortran` appropriate for your operating system).

(4) Edit `prepa.par` to compute the time interval 0 (`datefin`) to 100 (`datedebut`) Ma:

```
&NAMSTD
nomfich='ELL.BIN',
nomfichder='DER.BIN',
nomascpos='ORBELP.ASC',
nomascneg='ORBELN.ASC',
datedebut=&minus;100.D0,
datefin=0.D0,
statut='unknown'
&END
```

(5) Edit `integ.par` to compute the time interval 0 (`datefin`) to 100 (`datedebut`) Ma:

```
&NAMSTD
nomfich='ELL.BIN',
nomfichder='DER.BIN',
pas=200,
nechant=5,
datefin=0.D0,
datedebut=&minus;100.D0,
statut='unknown',
ecritpos='oui',
ecritneg='oui',
fichrespos='PRECP.ASC',
fichresneg='PRECN.ASC'
&END
```

(6) Edit `climavar.par` as follows:

```
&NAMSTD
nomascpos='ORBELP.ASC',
nomascneg='ORBELN.ASC',
nomprecpos='PRECP.ASC',
nomprecneg='PRECN.ASC',
nomsolpos='CLIVARP.ASC',
nomsolneg='CLIVARN.ASC',
datedebut=&minus;100.D0,
datefin=&minus;0.D0,
statut='unknown'
&END
```

(7)   Download the La2010a (or b, c, or d) orbital solution file `La2010a_alkhqp3L.dat` from:
http://www.imcce.fr/Equipes/ASD/insola/earth/La2010/index.html
Edit this file to delete columns 2 and 3 ("a" and "l"); save as `ORBELN.ASC`.
(Mac users will subsequently have to run the output file through `mac2unix`.)

(8)   Run `prepa`, then run `integall`, choosing `FGAM=1` and `CMAR=1.3` to match output of nominal La2004); finally, run `climavar`. The output file `CLIVARN.ASC` has four columns for `t`, `e`, `eps`, and `CP`, where `t`=time in kiloyears before present (negative), `e`=eccentricity, `eps`=obliquity (in radians), and `CP=climatic precession` (precession index). (Note: `FGAM` and `CMAR` can be varied to constrain geodynamical effects back through time, as was done in `Lourens et al., 2001`).

A.5   Other Key Resources

http://sourceforge.net/projects/cs-cmdtools/files/ (*mac2unix*)

http://cran.r-project.org/bin/macosx/tools/ (*gfortran*)

http://www.lsce.ipsl.fr/logiciels/ (*Analyseries: Mac OS only*)

http://www.atmos.ucla.edu/tcd/ssa/ (*SSA-MTM Toolkit: Linux or Mac OS*)

http://folk.uio.no/ohammer/past/ (*PAST: Paleontological Statistics, Windows only*)

http://paos.colorado.edu/research/wavelets/ (*Wavelet Software: FORTRAN, IDL, MATLAB*)

http://www.ncdc.noaa.gov/paleo/softlib/redfit/redfit.html (*REDFIT—Windows*)

References

Kodama, K. P., D. J. Anastasio, J. Pares, and L. A. Hinnov (2010), High-resolution rock magnetic cyclostratigraphy in an Eocene flysch, Spanish Pyrenees, *Geochemistry, Geophysics, Geosystems, 11*. DOI:10.1029/2010GC003069.

Laskar, J., F. Joutel, and F. Boudin (1993), Orbital, precessional and insolation quantities for the Earth from −20 Myr to +10 Myr, *Astronomy and Astrophysics, 270*, 522–533.

Glossary

Aliasing Low-frequency spectral peaks that do not represent real periodic behavior in a time series, but rather are an artifact of under-sampling the highest frequency cycles in the time series. Aliasing occurs when the sampling interval is greater than the interval necessary to sample the shortest period cycle twice per cycle.

Anhysteretic remanent magnetization (ARM) A laboratory remanence that results from applying a small DC magnetic field ($\sim$50–100 μT) to a sample in the presence of an alternating magnetic field that is decreased from some peak value to 0. In most laboratories, the peak field is about 100 mT. A partial ARM can be applied by only switching on the DC field over a limited range of alternating field during the decrease from a peak alternating field value.

ARM susceptibility (χ_{ARM}) In this case, it is not the susceptibility that causes an induced magnetization, but the ARM acquired as a function of the DC magnetic field used for the application of the ARM. Usually it is calculated by dividing the ARM by the strength of the DC field used during application of the ARM.

Astrochronology The astronomical calibration of cyclostratigraphy.

Band-pass filter A filter that passes a range of frequencies and suppresses or attenuates frequencies outside that range.

B_c The coercivity of a sample that results from a hysteresis loop measurement. It is the field that is required to reduce the magnetization of a sample to 0 while the field is being applied.

B_{cr} The coercivity of remanence of a sample that results from a hysteresis loop measurement. It is the field required to reduce the magnetization of a sample to 0 after the field has been turned off, i.e., the sample is measured in 0 field.

Chemical or crystallization remanent magnetization (CRM) The remanent, or spontaneous, magnetization acquired when a ferromagnetic mineral grows chemically, or crystallizes, in the presence of a magnetic field. When new crystal grains grow through a certain volume, typically submicron in size, the spontaneous magnetization becomes stable.

Chron A period of one dominant geomagnetic polarity, although shorter periods of opposite polarity (subchrons) can exist within a chron. The

Rock Magnetic Cyclostratigraphy, First Edition. Kenneth P. Kodama and Linda A. Hinnov.
© 2015 John Wiley & Sons, Ltd. Published 2015 by John Wiley & Sons, Ltd.

polarity chron numbering system is based on the numbering of seafloor magnetic anomalies up to ~157 Ma. The numbering system uses "C" as the designator for normal polarity anomalies back to the end of the Cretaceous Long Normal superchron at 83 Ma. Before that the designator is "M" for the Mesozoic chrons back until 157 Ma when the geomagnetic field reversed rapidly during the Late Jurassic. The subchron numbering system is complicated. See Tauxe (2010) for details.

Cyclostratigraphy The study of periodic behavior of the properties of a sequence of sedimentary rocks.

Demagnetization The procedure used to isolate the most ancient remanence in a rock sample. Less-magnetically stable components of magnetization are removed by a stepwise procedure using either alternating magnetic fields or temperatures below a magnetic mineral's Curie temperature, at which point all remanence is removed.

Diamagnetism The induced magnetization acquired antiparallel to the applied field by non-ferromagnetic particles.

Eccentricity The deviation of the Earth's elliptical orbit around the Sun from circularity. An eccentricity of 0 is a circular orbit, an eccentricity of 1 is a parabola. Earth's orbital eccentricity has ranged from 0.0034 to 0.058 over the past several hundred thousand years.

Evolutionary spectrogram A power spectrum that depicts the changes in the spectral peaks through the time series. It is calculated from a spectral estimator applied to a window moving through the time series.

Fast Fourier Transform (FFT) An algorithm that computes a discrete Fourier transform on a digital computer.

Ferromagnetism The permanent magnetism acquired by iron oxide and iron sulfide particles whose behavior is described by a hysteresis loop.

First-order reversal curve (FORC) A curve that results from contouring the second derivative of the surface fit to the magnetizations measured from multiple hysteresis loops with decreasing peak fields, and changing the coordinate system from that used for the generation of the hysteresis loops. The FORC diagram that results can be used to measure the coercivity of the collection of magnetic grains in a sample as well as the importance of magnetic interactions between the magnetic sample grains. FORC diagrams can characterize particle magnetization behavior.

Geomagnetic polarity time scale (GPTS) The record of the polarity intervals, reversed and normal, of the geomagnetic field as a function of age, through geological time. The GPTS is divided into chrons and subchrons and is based on the record of seafloor magnetic anomalies back to ~170 Ma.

Hard Isothermal Remanent Magnetization (HIRM) A rock magnetic parameter that quantifies the amount of "hard" or high coercivity magnetic minerals in a sample. It is calculated from: $HIRM = (SIRM + IRM_{-300\,mT})/2$, where SIRM is the saturation isothermal remanent magnetization and $IRM_{-300\,mT}$ is the IRM acquired in a backfield of 300 mT after a sample has been saturated.

Insolation The amount of incoming solar radiation received on a given surface area during a given time period.

Isothermal remanent magnetization (IRM) The magnetization that results from the application of a DC magnetic field to a sample.

J_{rs} Saturation remanence, also known as an SIRM. This is the greatest magnetization acquired by a sample after a sample has reached J_{sat}, but it is measured when the field that caused it has been turned off.

Low-pass filter A filter that only passes low frequencies and suppresses or attenuates all higher frequencies.

Magnetostratigraphy The polarity sequence recorded by either a section of sedimentary or igneous rocks. Typically, sedimentary rock sequences record the most complete and continuous records of geomagnetic field reversals. A local magnetostratigraphy of a rock sequence is then tied to the geomagnetic polarity time scale to assign absolute ages to the polarity interval boundaries.

Milankovitch cycles Changes in insolation caused by Earth's precession, obliquity, and orbital eccentricity.

Multidomain (MD) Magnetic mineral grains or particles which are large enough to have divided into several magnetic domains, i.e., regions which are uniformly magnetized.

Multitaper method (MTM) A modification of Fourier spectral analysis in which multiple spectral estimates are calculated from the same sample, each time the data series being multiplied by a different orthogonal taper function. The technique was developed by D.J. Thomson.

Natural remanent magnetization (NRM) A blanket term used to indicate all remanent magnetizations that are naturally acquired by a rock. It usually refers to the magnetization measured before a rock sample is demagnetized or before it has had a laboratory remanence applied to it.

Nyquist frequency The highest frequency (shortest period) cycle detectable by a spectral estimator for the given sampling interval. A cycle must be sampled at least twice. The Nyquist frequency is given by: $f_{Nyq} = 1/(2*SI)$ where SI is the sampling interval.

Obliquity The tilt of the Earth's rotation axis to the plane of the ecliptic, the orbital plane of the Earth around the Sun. It is currently about 23.4° and varies from about 22° to 24.5° with a 41 kyr period.

Paramagnetism The induced magnetization acquired parallel to the applied field by non-ferromagnetic particles.

Precession *or* **precession of the equinoxes** *or* **axial precession** The slow, continuous "wobble" of the Earth's rotation axis with respect to the stars. The precession of the Earth's rotation axis is caused by the gravitational pull of the Moon and the Sun on the Earth's equatorial bulge and causes a slow ~26 kyr precession of the Earth's rotation axis in space.

Precession index *or* **precession parameter** *or* **climatic precession** *or* **precession-eccentricity syndrome** The drift of the seasons and changes in Earth–Sun distance during the year is due to the Earth's precession and the precession of the orbital perihelion ("apsidal precession"). These motions produce a ~21 kyr cycle that is modulated by the orbital eccentricity and is a strong interannual insolation signal.

Precession of the orbital perihelion *or* **apsidal precession** The slow clockwise precession of the Earth's orbital perihelion in space.

Pseudo single domain (PSD) A small, multidomain ferromagnetic grain that behaves like a single domain grain. It is usually envisioned as containing just a few domains (regions of uniform magnetization), but PSD behavior could also be due to nonuniform magnetization configurations (vortex or flower) as indicated by micro-magnetic modeling.

Red noise A noise model for climate and geological time series in which low frequency noise has more power than higher frequency noise. The red noise spectrum results from the system having inertia or "memory."

Remanence Permanent, rather than induced, magnetization of a rock. Remanence is carried by ferromagnetic minerals. The important ferromagnetic minerals in rocks are magnetite, hematite, and some iron sulfides, such as greigite.

Saturation magnetization (J_{sat}) The ferromagnetic magnetization determined by a hysteresis loop measurement that is the greatest magnetization acquired by a sample in the presence of the magnetic field causing it.

Saturation isothermal remanent magnetization (SIRM) The isothermal remanence acquired by a sample after it has been exposed to a saturating DC field. See J_{rs} above.

Secular variation The natural directional and intensity variations of the geomagnetic field during a polarity interval. These changes occur over 10^3–10^4 year time periods.

Single domain (SD) Magnetic mineral grains that are small enough to have only one magnetic domain. The entire magnetic grain is uniformly magnetized.

Slepian tapers A specific kind of data taper used in an MTM spectral estimation. Also known as a discrete prolate spheroidal sequence or DPSS.

Smoothing Time-wise averaging of a time series to "smooth" out high-frequency variations.

S-ratio A magnetic mineral parameter that quantifies the ratio of low coercivity magnetic minerals (e.g., magnetite) to high coercivity magnetic minerals (e.g., hematite). The S ratio is calculated from $S = -IRM_{-300\,mT}/SIRM$.

Susceptibility The susceptibility is the proportionality constant between the induced magnetization and the field that is inducing it. $J = \chi\ H$. For AMS (anisotropy of magnetic susceptibility), the susceptibility is described by a second rank tensor (3×3 matrix).

Tapers A function that decreases to zero at each end gradually and smoothly. It is multiplied by a time series to remove the effects of sharp truncation at the ends of a finite length time series.

Virtual geomagnetic pole (VGP) The position of the north magnetic pole on the surface of the Earth that would generate the paleomagnetic direction observed for a rock sample assuming that the Earth's field is caused by a dipole at the center of the Earth. Since PSV (paleosecular variation) has not been time averaged, VGPs do not typically lie on the N geographic pole (spin axis of the Earth), but cluster around the N geographic pole during times of normal polarity.

Index

Note: Page references in *italics* refer to Figures; those in **bold** refer to Tables

Rock Magnetic Cyclostratigraphy, First Edition. Kenneth P. Kodama and Linda A. Hinnov.
© 2015 John Wiley & Sons, Ltd. Published 2015 by John Wiley & Sons, Ltd.

█████████████

Printed and bound by CPI Group (UK) Ltd, Croydon, CR0 4YY

27/10/2024

14580301-0005